ALLEN COUNTY PUBLIC LIBRARY

ACPL ITEM
DISCARDED

1. 25. 79

The Consumer's Guide to Video Tape Recording

The Consumer's Guide to Video Tape Recording

Boris Zmijewsky

STEIN AND DAY/*Publishers*/New York

First published in 1979
Copyright © 1979 by Boris Zmijewsky
All rights reserved
Designed by Ed Kaplin
Printed in the United States of America
Stein and Day/*Publishers*/Scarborough House
Briarcliff Manor, N.Y. 10510

Library of Congress Cataloging in Publication Data

Zmijewsky, Boris.
The consumer's guide to video tape recording.

1. Video tape recorders and recording. I. Title.
TK9960.Z58 778.59 78-7170
ISBN 0-8128-2535-7
ISBN 0-8128-6015-2 pbk.

This book is
dedicated to My Father, Pawlo Zmijewsky
(1906–1978)

2062138

Many Thanks To:

Art Ballant, Jr.
Ted Bonnitt
John Belton
Gary Knudsen
Paula Wooten
Karen Boghosian
Debra Zmijewsky
Steven Zmijewsky
Sony Corporation
JVC Corporation
Panasonic Corporation
Advent Corporation
G.E. Corporation
Phillips/MCA
Projection Systems, Inc.
Tentel Corporation

Special Thanks to

Jay O. Sturm
for his illustrations
and
Pauline O'Connell
for editorial work

Contents

1 • The Background to Today's Scene 1
2 • Types of Video Tape Recorders 7
3 • Consumer Comparison Guide to Video Tape Recorders 21
4 • Operator's Manual for a Video System 38
5 • Maintenance and Troubleshooting 65
6 • Tape Maintenance 84
7 • Video Beam Projector 91
8 • Video Disc 112
9 • Prerecorded Tapes 125
10 • The Camera 137
11 • Optional Equipment 142
12 • Film Production and Other Uses of VTR 154
Appendix: Companies Distributing Prerecorded Programs 169
Glossary 172

The Consumer's Guide to Video Tape Recording

1 • *The Background to Today's Scene*

The foundation of video tape recording was developed by German engineers during World War II. The magnetic tape recording technology that they invented was brought to the United States in the late 1940s. This magnetic technology evolved into the fundamentals of modern audio and video recording.

The first demonstration of an on-the-spot video machine at work took place in a Chicago theater in June, 1949. To the amazement of the audience, the first flashings on the screen were pictures of themselves, taken only a minute before. In 66 seconds, the image taken from TV had been processed and made ready for projection on screen via a standard piece of 35-millimeter optical equipment. One problem with this system was a lack of image clarity and poor sound quality. The audience did enjoy the transcription system when it showed them their own images almost as they watched, but this did not substantially increase theater attendance

At the onset of TV back in the late forties and up until 1949, members of The Society of Motion Picture Engineers looked at television as an unimportant colleague posing little threat to their dominance in public media.

However, as we all know, television did not remain in this subordinate position for very long. By 1949, over 675,000 TV set were located in the New York area alone, with another 55 million people estimated as within reach of programs. The production of TV sets reached as high as 118,000 in one month alone. On April 4, 1949, The Society of Motion Picture Engineers decided to change its name to include television engineers. TV was by now an economic threat to the motion picture business. To combat the threat, film companies thought up such promising new means as the lounge TV.

This was an optical barrel projector that shot TV shows onto a large TV screen 15 feet wide by 20 feet long. This may sound vaguely familiar in that it resembles the present-day Advent screen. This early idea of a barrel projector and large screen has been adapted for home use and is slowly becoming a threat to movies. The lounges gave those without a TV set a chance to see their favorite shows in a comfortable setting. The chief drawback to this idea was the admission charge for presentations that could be viewed at home for free. As a solo effort the plan failed, but when coupled with theaters it proved successful, and 20-inch TV sets were added in either the lobby or the lounge of already existing theaters. The Reade Theater chain had ten such television lounges installed in its theaters in the New York and New Jersey areas.

As the Advent screen had its start years ago, so did cable TV. Phonovision was the first cable TV system conceived and it was made available to homes by the telephone company at a fee. Motion pictures, along with other television material, were synchronized by wire and could be received by any home set. If not properly installed by the phone company, the reception was garbled beyond comprehension, but when installed correctly it gave a radiant broadcast. In order to receive the broadcast, the subscriber would call up the phone company and the show would be seen when it went on the air. The patron was billed for this extra service on his forthcoming phone bill. Also, in 1949, the Bell System added two video network channels between Philadelphia and Chicago, which made available three westbound and one eastbound channels.

Networks no longer had to share single-operation channels in each direction; TV was expanding by leaps and bounds. David Sarnoff, board chairman of RCA, told the thirtieth annual meeting of RCA stockholders in New York in 1948 that the outlook for TV was bright and it was "here to stay" because people liked TV and wanted it. It was continually gaining in popularity with the public; the advance of TV as a new industry could not be stopped.

The ultimate success of TV rested primarily upon programming. George Schutz, editor of *Better Theatres,* first conceived the idea of movies for TV in April,. 1949. Schutz felt that Hollywood should assume the task of producing full-scale films for home consumption, comparable to those of the motion picture industry, before someone else did. As we know, this did not materialize until much later. Instead, in June, 1949, United World, a Rank TV film subsidiary, was involved in the sale of seventy Rank films to TV stations. In the agreement the network acquired the pictures for a two to three-year period, during which they could be shown as often as the broadcaster deemed advisable. Prior to Rank's deal, Korda sold forty-two of his pre-1945 features to station KPIX, in New York.

On June 15, 1949, WPIX, owned by the New York *Daily News,* celebrated its first year with about 45 hours a week on the air. The station had 55 sponsors who bought up 34 percent of the station's program time and featured television films, presenting 24 Korda pictures as well as a Western series. The film industry found itself engaged in an intensive battle with TV. The Federal Communications Commission (FCC) attempted to keep film producers out of TV, not wanting them to monopolize the new form. The commission indicated that it would not disqualify a film company just because it was a film company. Nonetheless, it was not clear how much of a barrier film connections were in obtaining a TV license. To promote the mass appeal of TV the film industry produced diversified films that appealed to people in different age groups. No longer were films aimed at a blurred audience profile. Each film was now targeted for one of the specific interests of the viewers. There were science fiction films, westerns for kids, drama and mystery for adults, and films that attracted the teenagers.

Jack Keegan, general manager of the Paramount Northio Circuit, summed things up in 1949, stating that somewhere in every community is an audience for that "bad" picture or, in other words, almost any picture will have some appeal, if you reach the right audience. The film industry was fighting for its survival with the only means available–publicity. Local newspapers, billposters, and radio advertised the films and soon became everything to the film industry.

Today the motion picture industry is once again threatened, this time by cable TV, home video cassette machines, and video games. Of approximately 72 million households, 10.8 million have access to cable TV, and when the proportion reaches 30 percent it will rival commercial TV as well as the film industry. To counterract this new threat the film industry is focusing on films that lose impact when viewed on a small screen. Such films as *Star Wars* and *Close Encounters of the Third Kind* lose their vista and excitement when seen on a home screen. Sensoround is still another technique used to combat home viewing.

As the film industry was once hurt by TV, commercial television is now being threatened from within by such enemies as the cable system. Because of this threat, the television industry has been witnessing large-scale upheavals in management, programming, and strategy. Some members of the industry even go so far as to say that the network system will have only a limited existence.

Warner Cable's new two-way method of pay TV, the Qube tube, is the ultimate in pay cable, and may very well revolutionize TV. It is now being tested in Columbus, Ohio, where Warner Cable is operating a 26,000-subscriber system, with a 100,000-subscriber growth potential. The Qube is a two-part computer-terminal system allowing the subscriber to express opinions, interact in a classroom for credited college courses, and reject bad performances, as well as play games and win prizes. A black box near the TV set and the Qube console, connected to the set by a long wire, are the main components. The console offers 30 channels, 10 of which are premium-pay channels. The other 20 are devoted to 10 conventional TV channels and to 10 community channels.

Its computer system is the key to the Qube's outstanding performance. Every 6 seconds the computer scans each home terminal to note the channel being viewed and the response button being touched, and to check whether it is functioning properly. If you are tuned to any of the premium channels the computer will bill you automatically after two minutes of viewing. The Nielsen ratings will become obsolete as the Qube cable system catches on. Showtime as well as Home Box Office currently use this system. Recently, Fanfare TV and RCA signed an agreement for satellite distribution in the south and southwest, for movies as well as for special events.

Two major problems affecting the cable television industry today are "churn" and illegal converters used by nonsubscribers. "Churn" is the term used for subscriber discontent. Illegal converters can be stolen, imported from Canada, or, by the electronically handy, made at home and connected to a regular cable. Some people have been able to obtain pay TV by simply pushing a combination of buttons on a regular CATV converter. To counter this, the scrambling method was devised. This method transmits the picture in improper order or in synchronized signals which are descrambled in the subscriber's home. However, even this system can be foiled by the nonsubscriber with the purchase of an expensive descrambler. Two devices that counteract the illegal descramblers are the negative and the positive traps, which remove the pay TV picture from nonsubscribers' lines. The ultimate security system is the two-way closed cable system. In this system every subscriber, paying or nonpaying, is monitored by a head-end computer that gives a constant readout from each household, lists the channel it is tuned to, and either bills it for the show or disconnects it. One interesting finding that has come to light with the introduction of Qube is that people frequently change channels during commercial breaks.

However, the present growth of cable TV could be checked by the return of a new form of Phonavision. This revised system works on an ordinary phone, connected to the set by a single wire composed of optical fibers. Installed and maintained by AT&T, this wire has the capacity to carry an abundance of TV programs and all sorts of data communications, as well as ordinary phone messages. Such advanced technology might challenge, if not undo, cable TV altogether.

Today video tape recorders, too, are revolutionizing TV viewing. These machines enable viewers to rearrange a television schedule to fit their personal needs. Not only can the viewer watch one show while taping another, but with an automatic timer, he can tape a program while away from the set. And with the aid of the pause button, commercial interruptions are eliminated and the viewer can build a library of his favorite films and TV shows. The video tape recorder (VTR) system is easily installed and operated, and can even be played back on the 7-foot Advent screen. Home movies can be recorded on cassettes and viewed on TV. And, if you have the money to buy your own color video camera (or black-and-white, for a much smaller sum), you can escape commercial entertainment altogether by making your own movies—say, your own honeymoon?

Because of the current popularity of video cassette recording, cassettes are in great demand, and one often has to try several stores before obtaining any. Many people buy in quantity, some as many as twenty at a time. Cassettes come in two sizes (half-inch and three-quarter-inch). However, they are not interchangeable among all machines—a Sony tape will not fit an RCA player.

A legal battle in now taking place between Sony on the one hand, and Universal Studios and Disney, on the other, concerning the infringement of copyright laws. For many—and obvious—reasons, the studios only want video players in the home, with no recorders allowed. The top seven film producers (Fox, Columbia, United Artists, MGM, Avco, Embassy, and Warner Bros.) have also filed briefs in support of the plaintiffs. However, if Sony loses, there would appear to be no method of effective enforcement against already existing home recorders. TV executives are more concerned with the threat to prime time viewing and ratings than with copyright laws. Audience size sets the price for TV commercials, and the VTR systems have upset this balance by allowing viewers to watch shows at their own convenience rather than at that of the TV networks.

2 • *Types of Video Tape Recorders*

Video Tape Formats

The video recorder was once used only by large advertising agencies, the army, broadcasting companies, and schools. It can now be found in the home of the average housewife, taping her favorite soap operas.

There are two types of video tape recording formats currently in use: transverse (also called quad head), and helical (slant track). Both formats operate on a 30-frame-per-second time base.

Transverse

This type is the broadcast industry standard. It uses four rotating record heads to put the video signal on a 2-inch-wide tape. (Its nickname, "quad," means four, as in quadruped and similar words.) These heads rotate on an axis almost perpendicular to the direction of the tape transport, creating a head-to-tape speed of 1500 inches per minute. This is the head-to-tape velocity necessary to achieve satisfactory picture quality.

In addition to the transverse video tracks, three other tracks are laid longitudinally on the tape: the audio track, control track, and cue track.

Helical

A helical VTR is a tape machine in which the tape is wrapped around a head drum of large diameter; in traveling around the drum, the tape path takes the form of a helix, or a single turn of a spring. The tape leaves the drum at a level different from that at which it joined the drum, and it can be wrapped either halfway or completely around the drum.

A half-inch or three-quarter-inch video cassette machine is basically a helical video tape recorder (VTR), also called a video cassette recorder (VCR).

At the present time there are four basic types of video cassette machines available from manufacturers.

Playback Deck

This machine is simplest to operate. It is used only to play back recorded programs. It is small, light, and less expensive than the recording machine, but is unable to record or erase a program.

Video Record Deck

This is similar to the TV record deck (see below), capable of the same functions, but cannot tape off-the-air programs. It is used by small industrial or educational firms, to which off-the-air programs are not of major importance.

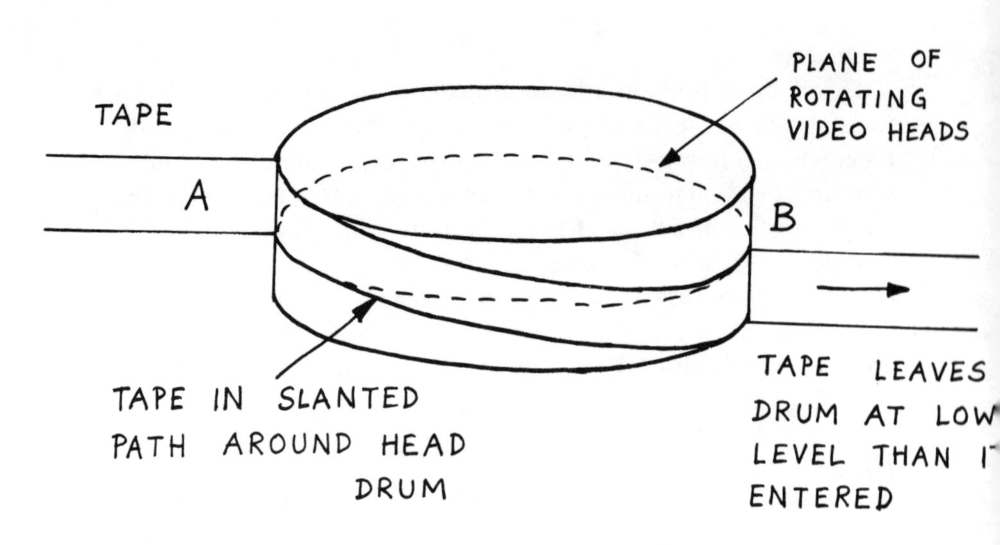

Tape path of a helical VTR

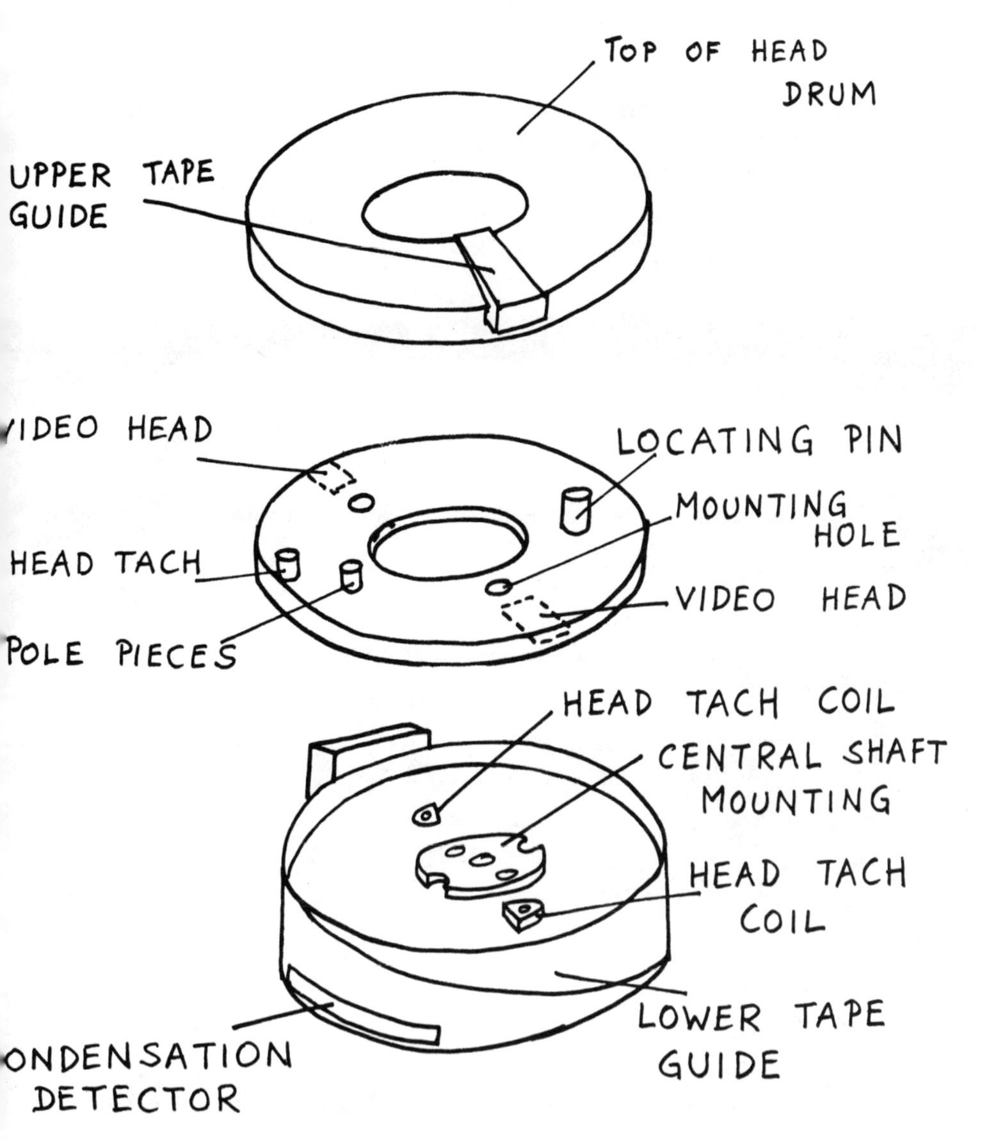

Exploded view of the head drum

TV Record Deck

This record machine is easily distinguished from the other decks: It is larger, because it includes a tuner at the right-hand side that permits recording of on-the-air programs. An antenna, connected to the rear terminal of the deck, allows a TV program to be received, just as with a normal TV set. The program can then be recorded directly onto tape and viewed with either a monitor or a TV set. The record machine has a continual electronic-to-electronic signal and radio-frequency output; therefore the program can be viewed both before and during recording.

Playback Deck: Sony VP-1000

The tape provides two audio tracks for recording, either separately or simultaneously, making postproduction dubbing onto one of the tracks possible. Separate playback of tracks is also possible, providing the capacity of stereo music or dual-language commentary.

Portapak

A portapak is a lightweight (20–30 pounds, depending on the model), portable video cassette recorder or reel-to-reel recorder. The recorder has a built-in capstan servo system and an internal crystal reference that maintains a stable, jitter-free tape speed. It also has a built-in color-processing circuit, thus eliminating the need for a separate color adapter, and allowing for full-color recording plus playback at any time. A dropout compensator and internal sync generator are also included. A switchable multipurpose meter indicates video and audio levels, battery power, and servo lock.

The portapaks are great for location shooting, and, most important, for industrial and commercial use.

Cassette Recordings and Playback

The broadcast and video format used in the United States is referred to as the North American Television Standards Commission (NTSC) system: 525 lines at 60 cycles per second (60 Hertz); that is, the electron beam sweeps a total of 525 horizontal lines to create the television picture. All video cassette machines will play back into a domestic color TV set or color monitor and produce a color picture, and the same goes for black-and-white (B/W) programs.

The first step is to make sure that the machine is plugged into the wall power outlet (120 volts VAC, 60 Hz). Then the power switch can be depressed and the green (or red) indicator light should come on, indicating it is safe to proceed. The TV set or monitor set can be plugged either into the wall outlet or into the auxiliary power outlet on the back of the machine.

The author shooting home movies with a portapak

Portapak: Sony AV-3400

Deck—Portapak, Sony AV-8400. Camera—AVC-3450

Portapak: Sony AV-8400

Panasonic portable VTR (UV-3085)
and camera (WV-3085)

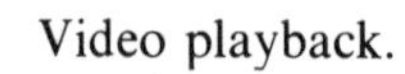

Video playback.

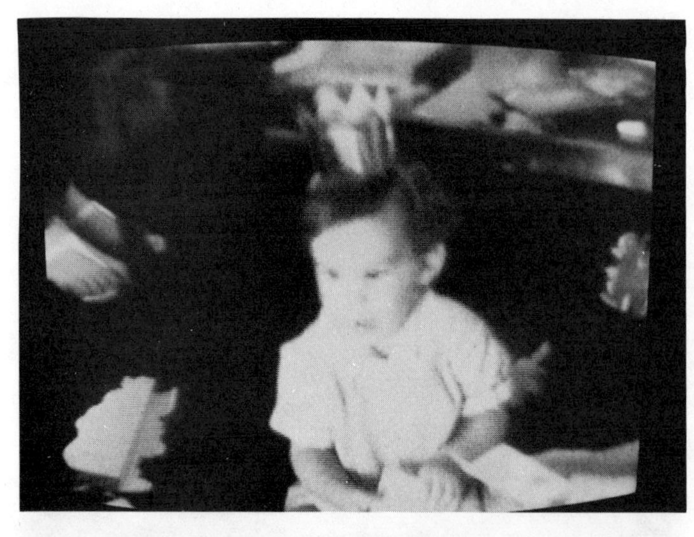

A perfect video playback.

The snowy appearance in the bottom portion of the picture is due to clogged heads. The video heads are dirt and must be cleaned.

The breakup in the pictur is due to a tracking error.

The General Controls

The general controls are clearly marked as to their location on each machine.

Power Switch

This must be used before any other function can take place. The power is turned on and off by hand. This switch is often a push on/push off button that is mounted away from the rest of the controls. Associated with it is an indicator light, usually green (or red), to show when the power is on.

Main Function Key (Play or Forward Key)

Depressing the play or forward button runs the tape out of the cassette and around the threading path and heads. A warning light then comes on to signal that the threading is in progress, and remains lit until the threading is complete. The play button stays down and the playback mode remains active. It is impossible to use any other key unless the stop button is first depressed.

Record

When this button is pressed it will stay down only when a cassette is inserted and the safety cap, or tab, is included. You will then receive a picture on your TV set or monitor. With no cassette inserted, or with the safety cap or tab omitted, the record button will not function. This button will not thread the tape.

To make a recording, the record button should be held firmly down and the play button pressed. This will lock the record button in place and cause the tape to be threaded. When the recording has ended, the stop button must be used to release the record and play buttons. This action will also unthread the machine.

As in all tape recording, the use of the record function will wipe away all previous material, including the video, the control track, and the audio track. It will also unthread the machine and reinsert the tape into the cassette, rewinding for about 10 seconds of program time.

Stop

This is a most important button. It must be used between all other modes of operation to stop whatever mode the machine is in.

Rewind and Fast Forward

Both of these functions of the tape are conducted with the tape inside the cassette.

Video Cassette Controls

TV Tuner

This TV tuner is exactly the same as that found in any domestic TV set and should be used as such. It has two outputs, a standard video signal and an audio signal. All machines with a tuner can record off-the-air programs in color and then play them back instantly into the same TV set.

Aft Switch

This switch gets automatic fine tuning into the circuit.

Pause Control

Lets you edit out any unwanted scenes during a recording, or stop the machine momentarily during playback. The pause mode will automatically be released after 7 minutes and the recorder will revert to the stop mode.

Memory Counter

To remember where your recorded program begins, simply set the memory switch and tape counter when you start your recording; when you are ready to play it back, you can rewind and it will stop automatically at the original starting point.

Audio Dubbing

This lets you replace a program's original sound with that from another source, such as a stereo, radio, or microphone.

Dew Indicator Lamp

If the machine is moved directly from a cold to a warm location, moisture may condense in internal parts. To prevent possible damage under this condition, the dew indicator lamp will light, indicating that excessive moisture has been detected within the set, and reject the cassette.

Excess moisture can cause the tape to stick. Keep the set's power on and wait until the moisture is evaporated by the built-in fan and heater: the indicator lamp will then go out.

If the lamp remains on for several hours, consult your dealer.

Headphone Jacks

Not all VTRs come with headphone jacks. Use headphones with an impedance of 80 ohms and equipped with a standard phone plug; when stereo headphones are used, nonaural sound will be heard in both headphones. For private listening, turn down the TV control.

Still Frame

This is usually found only on the better VTRs. The tape transport is stopped but the heads are allowed to rotate, producing a still-frame picture on the TV screen.

Erase Head

The erase head is energized in the record mode. Therefore, whenever a new recording is made, the previous recording on the tape is erased. Should you desire to erase a section of tape without recording, disconnect all the signal sources connected, and run the tape in the record mode. (Check the video cassette prior to loading to make sure that the red button is in place.)

RF Modulator

The purpose of this instrument is to allow a video cassette to be played back on a domestic TV set. The RF unit is installed within the compartment in the bottom or the back panel of the video deck. The RF unit is preset to an operating channel (3 or 4) inactive in the local area. If the set is moved to an area where the selected channel is in use, it will be necessary to reset the channel selector on the RF unit to another channel.

CAUTION:

Connection between the machine RF "out" terminal and the antenna terminals of the TV receiver should be made only as shown in your instruction manual. Failure to follow these instructions may result in operation that violates the regulations of the Federal Communication Commission regarding the use and operation of RF devices. Never connect the output of a machine to a receiving antenna or make simultaneous (paralleled) antenna and machine connections to the antenna terminals of a TV receiver without the use of an approved splitter.

Mistracking

A mistracking tape is caused by a rotating video head that scans two video tracks during one revolution. Mistracking is visible as a few lines of noise across the picture, which appear to run up or down the screen, causing a shuddering effect. To restore a perfect picture, the tracking control should be slowly turned until this effect disappears.

In some cases, streaks or snow may appear in the playback of tapes made on other VCRs. To correct this condition, turn the tracking control clockwise or counterclockwise until the best picture is obtained. When playback of this particular tape is finished, return the knob to the center detent position (Fix).

Hooking

A hooking or bending of the picture often appears at the top of the screen, and usually the image is not stable; the effect is often called "flaggers." It is caused by a difference between tension of the tape during playback and tension during recording. May factors can cause this, varying from bad tape to a maladjusted machine to atmospheric conditions. It is corrected by using the skew or tension control. Unfortunately, the range of this control is often inadequate to correct this fault.

Color Lock Adjustment

The video cassette recorder is preadjusted at the factory to produce correct color. Therefore, it is rare that an adjustment of the color lock will be required. However, if the picture should suddenly lose color or should not maintain correct hue, adjust the color lock control on the rear panel. Turn the control slowly clockwise or counterclockwise with a screwdriver until a normal color picture is restored. When playback of this particular tape is finished, return the screw to its center detent position. The control should be screwed back; if not, it will affect the next tape you play.

Capstan Servo

The capstan servo circuitry performs the following operations:

1. Keeps the speed constant.
2. Changes the tape speed when necessary.
3. Functions as the tracking servo during playback.

Dropout

Since video tape is a precision-made product of modern technology, it is vulnerable to many imperfections which can affect the working parts of the tape and the equipment.

Many of these resulting imperfections in the video image will remain unnoticed by the average viewer, but in time he can train his eye to recognize these distortions. The biggest problem facing tape viewers and users is dropout. Dropouts are seen as black or white holes in a picture, ranging from a tiny speck, hardly noticeable, to a mass of tiny specks resembling snowflakes. Dropout affects picture quality and sound. It is caused by momentary loss of head-to-tape contact. This is due to a hole in the oxide surface of the tape. The dirt and debris that settle on the tape not only affect image quality, but wear down the heads more rapidly, causing replacement more often than normally necessary.

3 • *Consumer Comparison Guide to Video Tape Recorders*

Characteristics of VCRs

Before purchasing a VCR, check each machine for the following features:

Mechanism

1. Cassette housing movement
2. Loading and unloading
3. Operation of each mode
4. (Stop, Play, F. F., Rewind)
5. Eject lock
6. Pause operation
7. Unusual noise

Mechanism Control

1. Auto stop at both ends of the tape
2. Repeat and search operation

Recording Function

1. Playback picture quality
2. Switching point and tracking
3. Editing
4. Audio recording and playback

Others

1. Antenna select switch operation
2. Audio select switch operation
3. RF output
4. Remote control
5. Built-in tuner
6. Timer recording

How to Read the Specifications

Before you decide on a video cassette recorder, you must understand the specifications of each machine. From these you can judge for yourself the quality of the machine you might be interested in. A VTR is only as good as its specifications.

1. *Video recording signal:* What kinds of recording heads are being used.
2. *Video signal system:* Electronic Industries Association (EIA) determines audio and video standards in the United States.
3. *Operating temperature:* Self-explanatory.
4. *RF Output signal:* To which channel (3 or 4) you must set your channel selector to receive your picture on playback.
5. *Signal-to-noise (S/N) Ratio:* This is a very important specification to know before you buy your VCR. The higher number of dBs, the better your picture quality will be.
6. *Tape speed:* What speed the video cassette is traveling at. The higher the speed, the better video and audio you will receive.
7. *Horizontal resolution:* See Glossary.
8. *Frequency response:* See Glossary.

U-Matic Systems

Sony U-Matic VO-2600: Technical Specifications

General

Video recording signal–Rotary two-head helical scan system; luminance: fm recording; color signal: converted subcarrier direct recording

Video signal system–EIA Standards, NTSC color signals

Storage temperature– -10°C to +60°C (14°F to 104°F)

Operating temperature– -5°C to 40°C (41°F to 104°F)

RF output signal–Channel 3 or channel 4 (with optional RF kit) 75 ohms, unbalanced

VCR Characteristics

Brand	*Speeds*	*Audio Dub*	*Freeze Frame*	*Video S/N*	*Audio S/N*	*Weight*	*Power Used*
VHS FORMAT							
Curtis Mathes C-78	SP, LP	Yes	No	40	42	38 lbs.	45W
GE Command Performance 9000	SP, LP	Yes	Yes	43	40	30 lbs.	29W
JVC Vidstar HR-3600	SP, LP	Yes	Yes	45	40	31 lbs.	33W
Magnavox 8220	SP, LP	Yes	No	45	42	45 lbs.	38W
Panasonic PV-1100	SP, LP	Yes	No	42	42	38 lbs.	45W
Philco V-1000	SP, LP	Yes	No	43	40	38 lbs.	45W
Quasar VH-5010	SP, LP	Yes	No	42	42	38 lbs.	45W
RCA Selectavision VCT-400	SP, LP	Yes	No	40	42	38 lbs.	40W
BETA FORMAT							
Sanyo VTC-9100A	X2	No	No	43	40	44 lbs.	65W
Sears 5305	X2	No	No	43	40	49 lbs.	70W
Sony SL-8600	X2	No	No	45	40	37 lbs.	80W
Toshiba V-5310	X2	Yes	No	42	42	43 lbs.	65W
Zenith KR-9000W	X2	No	No	40	40	38 lbs.	65W

Power requirement–120V AC + 12V, 60 Hz + 0.5% (VO-2600); 120V AC + 12V, 50 Hz + 0.5% (VO-2600E)
Power consumption–130W
Weight–31.5 kg (69 lbs. 7 oz.)
Dimensions–591 mm (W) × 227 mm (H) × 417 mm (D) (23⅜″ × 9″ × 16½″)

Video

Input–1.0V (p-p) + 1.0V (p-p) -0.5V (p-p) 75 ohms, sync negative, unbalanced
Horizontal resolution–Monochrome: 320 lines; color: 240 lines
Signal-to-noise ratio–Better than 45 dB

Audio

Input (both channels)–LINE IN jacks: 10 dB, 100 kilohms, unbalanced; MIC IN jacks: -60 dB, suitable for microphones with 600-ohm impedance, unbalanced
Input–LINE OUT jacks (both channels): -5 dB (at 100 kilohm load), unbalanced; AUDIO MONITOR jack: 5 dB (at 100 kilohm load), unbalanced; HEADPHONES jack: 8 ohms
Frequency response–50 to 15,000 Hz
Signal-to-noise ratio–Better than 45 dB (both channels, 3% distortion at 1 kHz)

Tape Transport

Tape speed–9.53 cm/sec. 3¾ ips)
Wow and flutter–Less than 0.2% rms
Recording or playback time–60 min. (for Sony KCA-60, KC-60); 30 min. (for Sony KCA-30, KC-30)
Fast forward time–Within 6 min. (for Sony KCA-60, KC-60)
Rewind time–Within 4 min. (for Sony KCA-60, KC-60)

Special Functions

Pause mode–A still picture is displayed in the pause mode with Sony KCA or KCS type video cassettes.
Timer operation–Possible (with supplied timer recording adaptor)
Remote control–Possible (with Sony model RM-420 remote control)

Accessories Supplied

Video cassette tape KCA-10; timer recording adaptor FA-20; dust cover

Optional Accessories

Sony model RM-430 or RM-410 remote control; Sony model TT-100 tuner/timer; Sony model RFK-203FW or RFK-204FW RF kit.

Panasonic NV-¾″ Format Series Performance Data and Specifications

General

Power Source—120V AC, 60 Hz

Power consumption—NV-9100: Approx. 110 watts; NV-9300: Approx. 120 watts

Television system—EIA standard (525 lines, 60 fields); NTSC-type color signal

Video recording system—Two rotary heads, helical scanning system

Luminance signal—Frequency modulation recording

Color signal—Converted subcarrier direct recording

Audio track—2 tracks

Tape Transport

Tape speed—3¾ ips (95.3 mm/s)

Relative head-to-tape speed—404.3 ips (10.26 m/s)

Record/playback time—60 min. using NV-P26 tape; length approx. 1175 ft.

Fast forward/Rewind time—Less than 3 min. with NV-P26

Heads

Video—2 rotary HPF heads

Audio/Control—1 stationary head

Erase (NV-9300 only)—1 full track; 1 for audio dubbing

Input

Video Input (NV-9300 only)—LINE: 1.0V (p-p), 75 ohms, un-

The inside of a ¾″ U-Matic Sony VO-2600

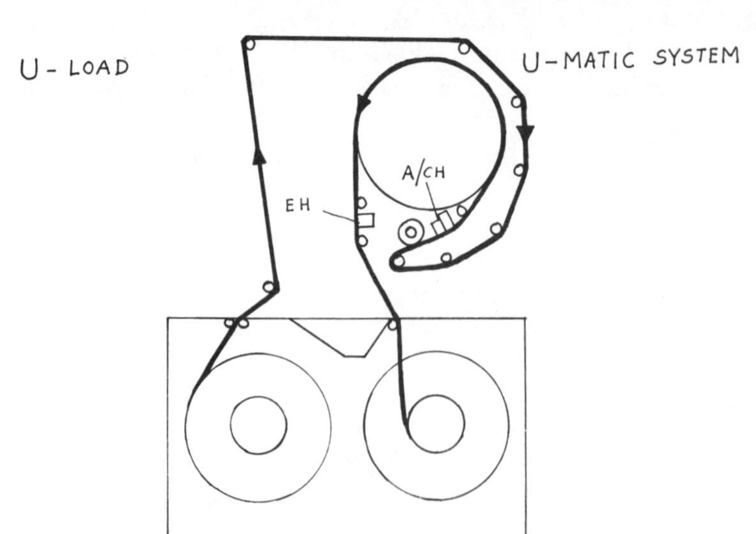

Tape threading on a U-Matic

balanced, UHF connector; TV: 1.0V (p-p), 75 ohms, unbalanced
8-pin connector
Antenna input/output—VHF: 75 ohms, unbalanced, F-type connectors; UHF: 300 ohms, unbalanced screw terminals (NV-9300 only)
Audio input (NV-9300 only)—LINE: -20 dB, 100 kilohms, unbalanced, × 2 RCA-type phono connectors; MIC: -78 dB, 250 ohms, unbalanced, × 2 phone jacks; TV: -20 dB, 100 kilohms, unbalanced, 8-pin connector (channel 2 only)

Output
Video output—1.0V (p-p), 75 ohms, unbalanced, UHF and 8-pin connectors
Audio output—LINE: 6 dB, 600 ohms, unbalanced, × 2 RCA-type phono connectors; TV: dB, 5 kilohms, unbalanced, 8-pin connector (channel 1, channel 2, or mix); HEADPHONE: -32 dB, 8 ohms, unbalanced, RCA-type phone connector (channel 1, channel 2 or mix)

Other
Horizontal resolution—B/W: more than 330 lines
(Monoscope test pattern)—Color: more than 240 lines
Audio frequency response—50 to 15,000 Hz
Signal-to-noise ratio—Luminance: more than 45 dB Rhode and Schwarz noise meter); Audio: more than 45 dB
Audio crosstalk—More than 40 dB
Operating temperature—45°F to 104°F (5°C to 40°C)
Operating humidty—Less than 80%
Weight (approx.)—NV-9100: 64 lbs. (29 kg); NV-9300: 73 lbs. (33 kg)

Dimensions (Approx.)
NV-9100: 21¼″ (W) × 9⅜″ (H) × 18 ⅝″ (D); 540 (W) × 238 (H) × 472 (D) mm; NV-9300: 24½″ (W) × 9⅜″ (H) × 18⅝″ (D); 621 (W) × 238 (H) × 472 (D) mm

Standard accessories

1 pc. AC power cord, VJA 33; 1 pc. VTR/ Monitor connection cable, NV-C15; 1 pc. dust cover

Optional accessories

¾" video cassette tape–NV-P23, 30 min.; NV-P26, 60 min. Head cleaning tape–NV-P20

Remote control unit–NV-AL52,

RF modulator–NV-U105, for TV channel 5; NV-U106, for TV channel 6

Sony U-Matic Type II–*VP-2260*
Features and Specifications

Power requirements–120 + 12V 60Hz + 0.5%

Power consumption–130W

Video signal–output: 1.0 + 0.2V, 75 ohms; horizontal resolution: Mono: 330 lines; color: 260 lines; signal-to-noise ratio: more than 45 dB; output: Line-5dB (2 kilohms); Monitor: 5dB (2 kilohms); signal-to-noise ratio: more than 46 dB; frequency response: 50–15,000 Hz; 3¾ ips

Record/Playback time–60 min. (for KC/KCA-60 tape)

Wow and flutter–Less than 0.2%

Operational position–Horizontal (+ 30°)

Fast forward time–Less than 6 min. (for KC/KCA-60 tape)

Rewind time–Less than 4 min. (for KC/KCA-60 tape)

Temperature–Storage +14° to +140°F/Operation +41° to +104°F

Dimensions–20⅜" (W) × 9" (H) × 16⅜" (D)

Weight–59.5 lbs.

Accessories supplied–Dust cover/instruction manual

Betamax System

The Betamax video recorder is a small-sized video cassette system similar to the three-quarter-inch U-Matic. The difference between them is in the size of the tape cassettes and the weights of the machines. The Betamax recorder is especially designed for home use, while the three-quarter-inch U-Matic is made for commercial broadcast use. Recording and playback of a TV program can be done by connection to an ordinary television set. All the accessories—buttons and the like—that were usual in earlier video machines are dispensed with, making it as simple to use as an ordinary audio cassette recorder. The Betamax is a lightweight, compact, low-cost machine, which, together with the extremely low cost of the tape, makes the unit ideal for the home consumer.

Here are its main characteristics:

Recording time is now 2 hours, double that of earlier models.

During playback, tape speed changes automatically to the tape speed that was used during recording.

Tape economy has been improved.

Counter memory switch is provided to allow the machine to rewind to a preselected point on the tape.

Play button actuates the automatic antenna selection mechanism to feed playback signals to the receiver.

In all modes, except playback, the incoming signal from the antenna is displayed directly on the television screen

Inputs for video camera and microphone are provided.

Recordings are initiated by depressing only one button.

Level adjustments of video and audio signals are performed automatically during recording.

Automatic color-B/W signal selection circuit senses the presence of color in the signal being recorded and sets up proper signal processing automatically.

Pause mechanism is provided for stopping the machine momentarily during recording or playback.

Built-in tuner permits one program to be viewed while another is recorded.

Timer permits recordings to be made while the machine is unattended.

Betamax stops automatically at the end of the tape.

B - LOAD

BETA - SYSTEM

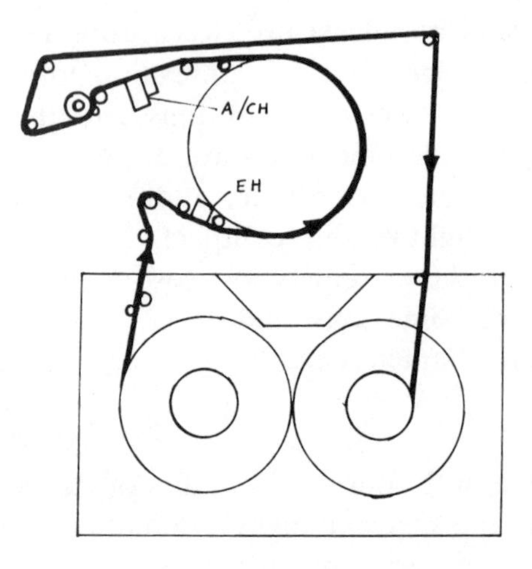

Tape threading on a Beta

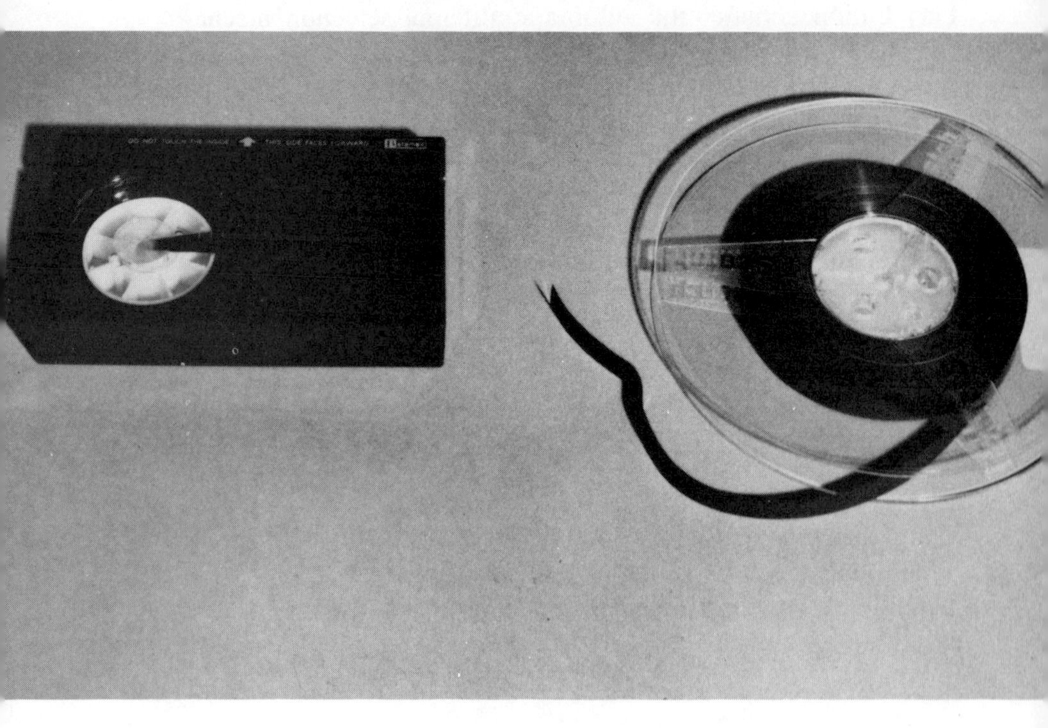

Betamax cassette vs. reel-to-reel tape

Betamax Operation

The Betamax is a very simple system to operate. The video cassette is inserted horizontally into the front-loading carriage compartment slot, which is then pressed down. This causes the tape to be automatically threaded, where it will rest until one of the operation buttons (play, record, rewind, fast forward) is pressed. To remove the cassette, the eject button is used. This automatically unthreads the tape, retracts it into the tape cassette, raises the carriage compartment, and shuts down the main motor. Then the cassette is removed by hand.

TV sections are included in many models. The three control knobs are a VHF selector, a UHF selector, and an automatic fine tuning button. Connecting an antenna to the back of the machine gives it the capability to pick up and record any station in the area, and allow the program to be viewed on a home TV set.

Half-Inch Beta Format–Sony SL-8600
General Specifications

General
Video recording system–Rotary two-head helical scan system
Video signal–EIA standards, NTSC color
Storage temperature– -20°C to +65°C (-40°F to +149°F)
Operating temperature– 5°C to 40°C (41°F to 104°F)
Antenna–75-ohm external antenna terminal for VHF 300-ohm external antenna terminals for UHF
Channel coverage–VHR channels 2-13; UHF channels 14-83
VHF output signal–Channel 3 or channel 4 (switchable), 75 ohms, unbalanced
Power requirements–120V AC +10%, 60 Hz +0.5%
Power consumption–80W
Weight–37 lbs.
Dimensions–520 mm (W) × 205 mm (H) × 410 mm (D); (20½″ × 8$\frac{1}{16}$″ × 16⅛″, w/h/d)

Video

Output–1.0V (p-p) +0.1 (p-p), 75 ohms, unbalanced, sync negative

Horizontal resolution–Monochrome; more than 280 lines; color: more than 240 lines

Signal-to-noise ratio–Better than 45 dB

Audio

Output–Less than 10 kilohms, -5 dB (100-kilohm load), unbalanced

Frequency response– 50 to 10,000 Hz (recording time selector 1 hr.); 50 to 8000 Hz (recording time selector 2 hrs.)

Signal-to-noise ratio–Better than 42 dB (recording time selector 1 hr.); better than 40 dB (recording time selector 2 hrs.)

Audio distortion–Less than 3% (at 333 Hz)

Tape Transport

Tape speed–4.0 cm/sec. and 2.0 cm/sec.

Maximum recording time–120 min. (with Sony L-500 Video cassette)

Fast forward time–Within 3½ min. (L-500)

Rewind time–Within 3½ min. (L-500)

Special Features

Pause mode

Accessories Supplied

Betamax video cassette tape; external antenna connector EAC-20W (75 ohm to 300-ohm matching transformer): external antenna connector EAC-22 (300 ohm to 75-ohm matching transformer): 75-ohm coaxial cable (2.5 M); dust cover

Optional Accessories

Cassette tape L-500/L-250; Betamax stand SU-61; Digital timer DT-30; TV/Betamax stand KVS-41/KVS-42; B/W camera AVC-1420; extension cable (25′) RFC-8; automatic cassette changer; camera cable VMC-3L; AG-120; all-channel splitter Y-100

Sanyo Betacord System
Specification–Model VTC 9100

Television signal–U.S. standard (NTSC) 525 lines, 30 frames/60 fields/sec.; color and B/W
Recording system–Rotary 2-head helical scan with azimuth recording
Tape speed–20 mm (0.79 ips)
Cassette length–1, 2, or 3 hours
Cassette dimensions–6⅛″ (L) × 3¾″ (W) × 1″ (H)
Rewind/fast forward time–Less than 4 min. for 2 hour cassette
Load/unload time–Less than 3 seconds
Video input–0.5–2V (p-p), 75 ohms, negative sync.
Video output–I.0V (p-p), 75 ohms, negative sync.
Video S/N ratio–Luminance better than 43 dB; chrominance better than 35 dB
Horizontal resolution–Color: 240 lines minimum; B/W: 250 lines minimum
Microphone input– -60 dB (600 ohms unbalanced)
Audio output–560mV (10 kilohms, unbalanced); phono jack
Audio bandwidth–50 to 7000 Hz +3, -4.5 dB
Audio distortion–Less than 2%
Audio S/N ratio–Better than 40 dB
RF adaptor–Plug-in channel 3 or 4
Power–117V AC, 60 Hz, 65 watts
Dimensions–19.5″ (W) × 14.6″ (D) × 7.7″ (H)
Weight–44 lbs.

Zenith Betatape System–JR9000W

Audio

Input–AUDIO IN: mini jack 100 kilohms, 10 dB MIC: mini jack; -60 dB, suitable for microphone with 600-ohm impedance
Output–AUDIO OUT: Mini jack, less than 10 kil ohms, -5 dB (100 kil ohm load), unbalanced

Frequency response—50 to 10,000 Hz with record time selector on (●); 50 to 8,000 Hz with record time selector on (●●)
Signal-to-noise ratio— Better than 42 dB, with record time selector on (●); Better than 40 dB, with record time selector on (●●)
Audio distortion—Less than 3% at 400 Hz, with record time selector on (●); Less than 4% at 400 Hz, with record time selector on (●●)

Tape Transport
Tape speed—4.0 cm/sec., with record time selector on (●); 2.0 cm/sec., with record time selector on (●●)
Maximum recording time—120 min. (with L-500 cassette);
Fast forward time—Within 3½ min. (L-500)
Rewind time—Within 3½ min. (L-500)

Accessories Supplied
RF unit RFU-534Z; video cassette tape; external antenna connector EAC-20Z (75 ohm to 300-ohm matching transformer); external antenna connector EAC-22Z (300 ohm to 75-ohm matching transformer); 75-ohm coaxial cable with F-type connectors (2m); switch stopper; dust cover

Optional Accessory
All-channel splitter

Video Home Systems (VHS)
Half-Inch VHS Panasonic Format—NV-8310

Power source—120V AC, 60 Hz
Power consumption—Approx. 29 watts
Television system—EIA Standard (525 lines, 60 fields); NTSC color signal
Video recording system—2 rotary heads, helical scanning sys-

tem; luminance: FM azimuth recording; color signal: converted sub-carrier phase shift recording
Audio track—1 track
Tape format—Tape width ½ inch (12.7 mm), high-density tape
Tape speed—1-5/16 ips (33.35 mm/s)
Record/Playback time—120 min, with NV-T120
FF/REW time—Less than 4.5 min. with NV-120
Heads—Video: 2 rotary heads; Audio/control: 1 stationary head; erase: 1 full track erase, 1 audio track erase for audio dubbing

2062138

M - LOAD VHS - SYSTEM

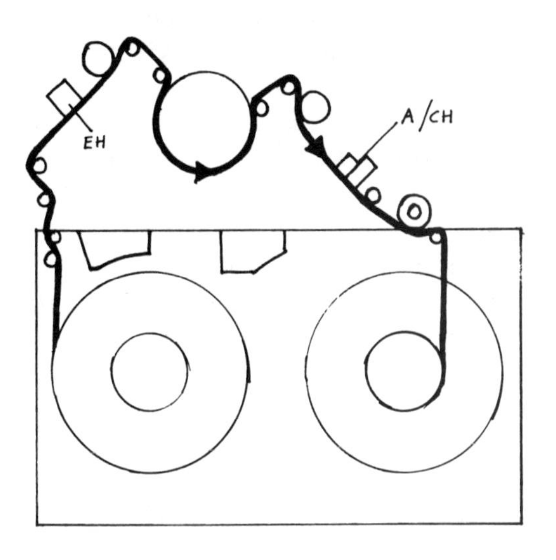

Tape threading on a VHS system

Input level

Video–TV Monitor Connector (8 pins) 1.0V (p-p), 75 ohms unbalanced; VIDEO IN connector (BNC) 1.0V (p-p), 75 ohms unbalanced

Audio–TV Monitor connector (8 pins) -20 dB, 100 K, unbalanced; MIC in jack -70 dB, 600 unbalanced; LINE IN jack (RCA) -20 dB, 100 K, unbalanced

TV tuners–VHF input ch 2, ch. 13; 75 unbalanced; UHF input ch. 14–ch. 83; 300 unbalanced

Output level–Video: TV monitor connector (8 pins); 1.0V (p-p), 75 unbalanced, VIDEO OUT connector (BNC); 1.0V (p-p), 75 unbalanced; audio: TV monitor connector (8 pins); 0.5 dB, 600 unbalanced; LINE OUT jack (RCA); -6 dB, 600 unbalanced; RF modulated: channel 3, 4, 5, or 6 73 dB (open voltage) 75 unbalanced

Video horizontal resolution –Color: more than 240 lines (on monoscope test pattern). B/W more than 300 lines

Audio frequency response–50 to 10,000 Hz

Signal-to-noise ratio–Video: better than 45 dB (Rohde and Schwarz noise meter); Audio: better than 43 dB

Operating temperature–41°F–104°F (5°C–40°C)

Operating humidity–35%–80%

Weight–36.3 lbs. (16.5 kg)

Dimensions–19⅛″ (W) × 16 5/16″ (D) × 7 5/16″ (H); 485 mm (W) × 414 mm (D) × 178 mm (H)

Accessories supplied–1 pc. Video cassette tape, NV-T60; 1 pc. 75-300 transformer, VSQ0015; 1 pc. 300-75 transformer, VSQ0057; 1 pc. F-F coaxial cable VJA0103; 1 pc. twin lead wire, VJA0102; 1 pc F connector, VSQ0051; 1 pc. dust cover, VYC0023

Optional accessories–½″ video cassette tape: NV-T120 approx. 810 ft., 120 min.; NV-T60 approx 417 ft., 60 min.; NV-T30 approx. 210 ft., 30 min.; RF converter, plug-in type; NN-U134/ch. 3 or ch. 4 switchable; NV-U105/ch. 5 or NV-U106/ch.6; Remote pause control, NV-A181

JVD Vidstar–VHS System–HR-3600 AV
VHS Standard Format

Recording system–Rotary, slant azimuth two-head helical scan system
Video signal system–NTSC-type color signal
Tape width–12.7 mm (½ in.)
Tape speed–33.35 mm/sec (1.32 inch/sec.)
Maximum recording time–120 min. (with JVC-T-120 video cassette)
Power requirement–120V AC, 60 Hz, 33 watts (40 watts with dew protection circuits activated)
Temperature
  Operating–5°C to 40°C (41°F to 140°F)
  Storage– -20°C to 60°C (-4°F to 140°F)
Video
  Input–0.5V to 2.0V (p-p), 75 ohms, unbalanced
  Output– 1.0V (p-p), 75 ohms, unbalanced
  Signal-to-noise ratio–More than 45 dB (Rhode and Schwarz noise meter)
  Horizontal resolution–More than 240 lines (color mode)
Audio
  Input–Mic: -67 dB; 10 kilohms, unbalanced; line: -20 dB; 50 kilohms unbalanced
  Output–0 dB, high impedance load
  Output impedance–1 kilohm, unbalanced
  Signal-to-noise ratio–More than 40 dB
  Frequency response–50 to 10,000 Hz
Dimensions–453 mm (17-⅞″) W × 147 mm (5-13/16″) H × 337 mm (13-5/16″) D
Weight–14 kg (31 lbs.)
Accessories provided–Power cord; video cassette tape T-30; dust cover; antenna cable; instruction booklet

4 • *Operator's Manual for a Video System*

The TV Set

The most recent television sets have a switchable VHF input. This input can be switched to 75 ohms or 300 ohms. The typical VHF input section on a TV set is shown on the accompanying illustration.

Some other TV sets may have a small moving bar or pin to select 75 ohms or 300 ohms. In any case, select 75 ohms for use with your VCR.

If the TV set does not have a 75-ohm VHF input, a matching transformer will be necessary. This matching transformer should be connected as shown in the illustration.

The VCR has built-in UHF and VHF tuners and signal splitters. Its connection and use with an antenna (or CATV cable) and TV set are very simple, and should be permanent. Normally, the UHF antenna (if used) and the VHF antenna (or CATV cable) are connected directly to the TV set.

When the recorder is used, the UHF antenna and the VHF antenna are connected directly to the recorder; the antenna signals then come out of the recorder and are connected to the TV set. The tuners on the TV do not affect the tuners on the recorders.

Antenna Connections

To make direct connections to the antenna terminals, strip the plastic from the end of the lead-in wire. Loosen the nuts on the terminals and wind the bare copper leads around the terminals. Make sure that the wires do not touch each other. Tighten the nuts.

Plug-in connections to the VHF or UHF antenna terminals using a conventional ribbon-type lead-in wire can be conveniently made with the EAC-12.

When using a 75-ohm coaxial cable for VHF reception, follow the instructions given in the illustration.

The EAC-13W permits convenient plug-in connection to the VHF or UHF antenna terminals when a 75-ohm coaxial cable is used.

Connection to a Cable-TV (CATV) System

Before operating the recorder with a cable-TV system, set the RF unit of the recorder to channel 3 or 4, whichever is not active in your area. If both channels are viewable, check which gives better results by switching the RF unit between channels 3 and 4.

In the configuration shown in the drawing, you can record programs from all CATV channels as well as VHF channels 2 through 13.

Set the TV channel selector to that of the RF unit. Set the channel selector on the recorder to the output channel of the converter. Set the Program Select or the recorder to the cassette position. Thus, the channel to be viewed or to be recorded is selected on the converter.

Tape Path in the Forward Mode

Once the forward mode is pressed, the tape from the supply reel table assembly first passes the guide and the tape-slack preventing plate inside the cassette. The tape-slack preventing plate applies braking pressure to the tape when slack occurs inside the cassette. This is necessary because the tape would be damaged by the cassette lid if slack tape were to come out of

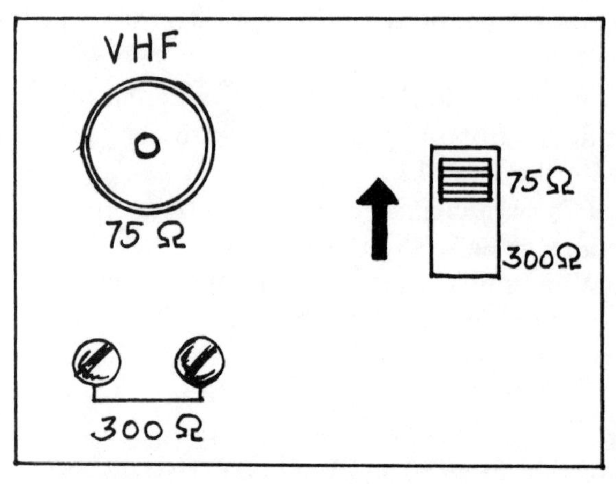

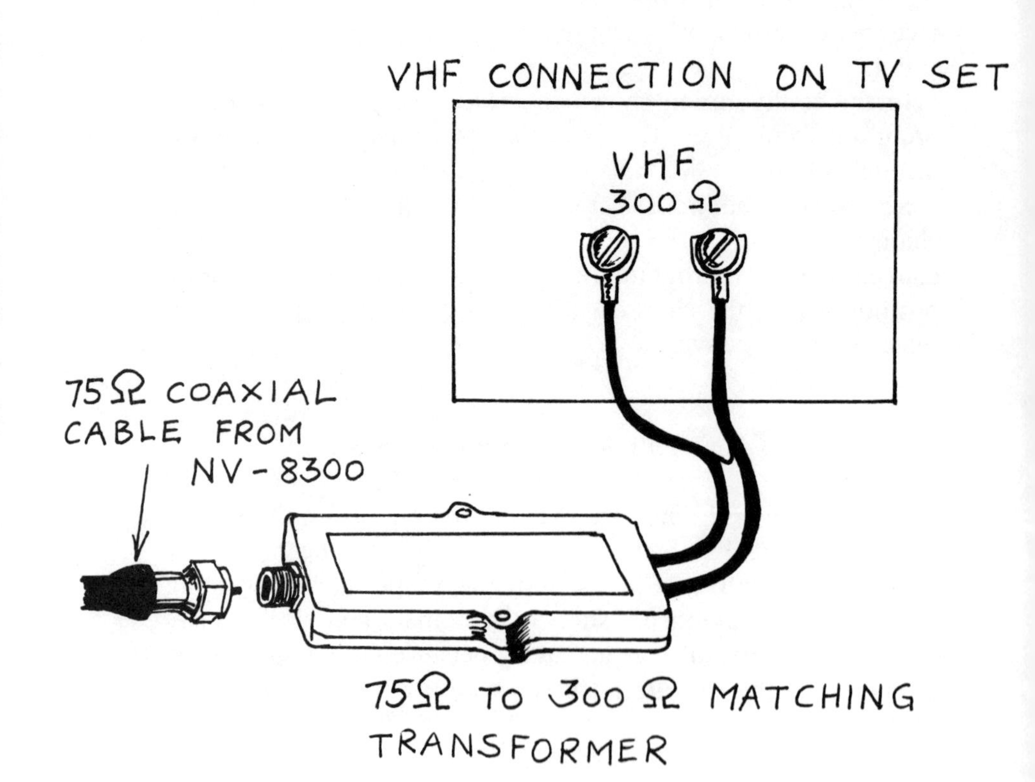

A transformer

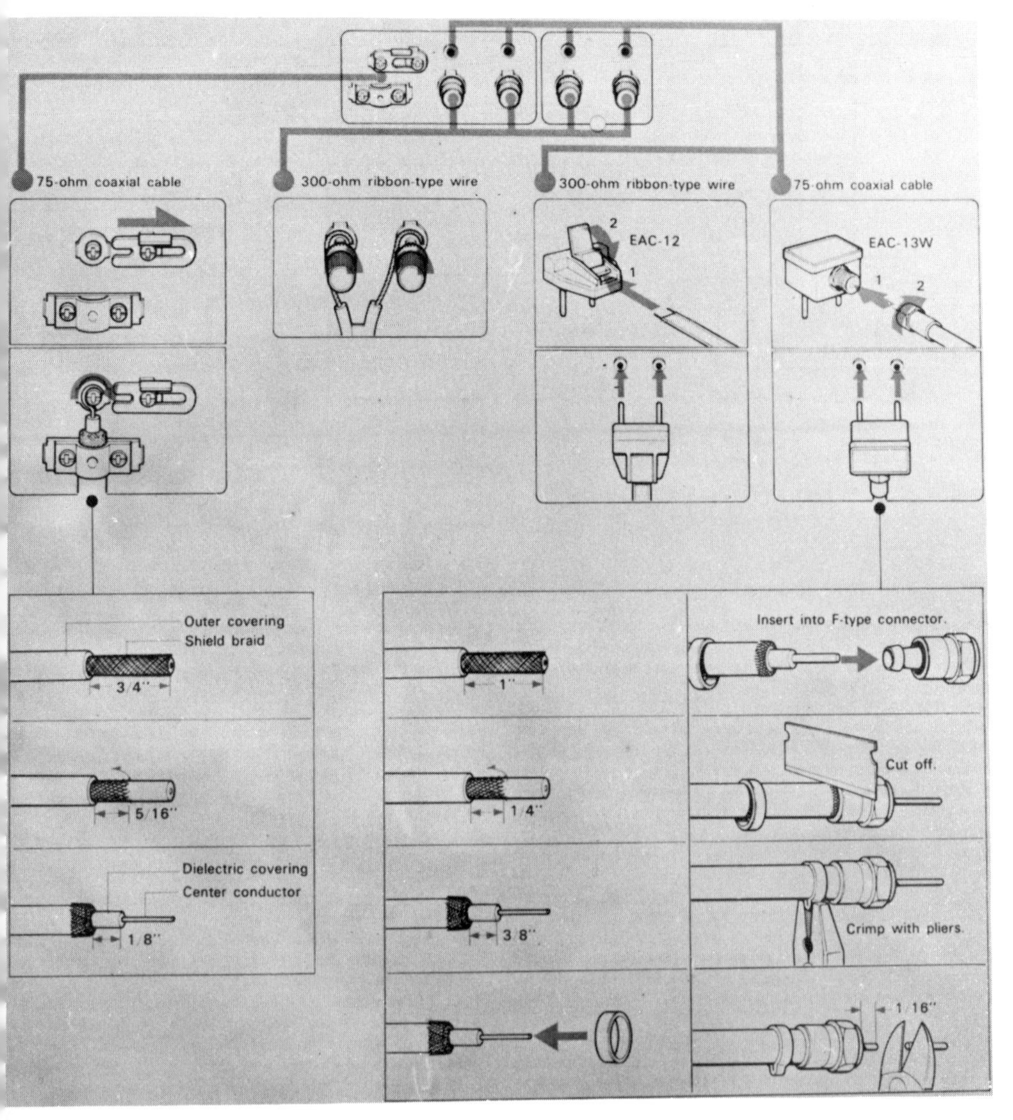

Different types of antenna connections

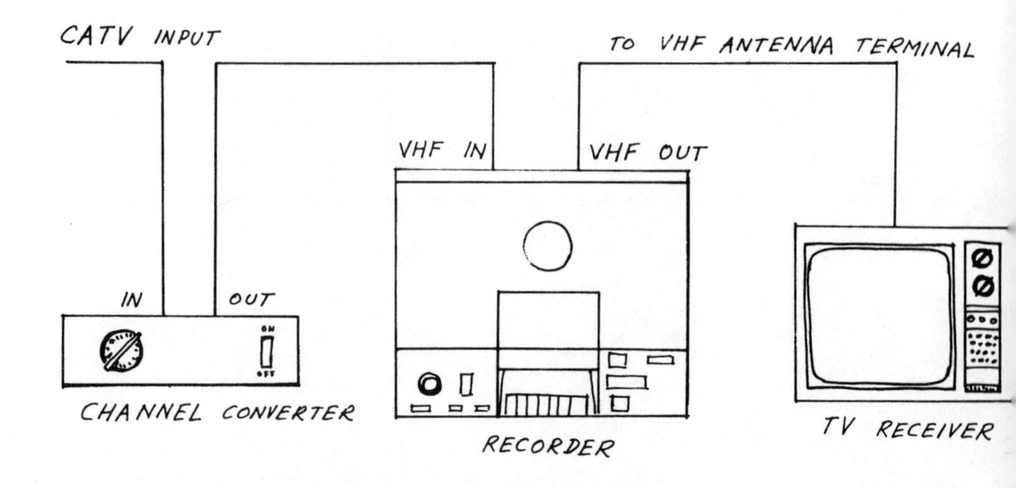

Connections to a cable TV system with the channel converter first

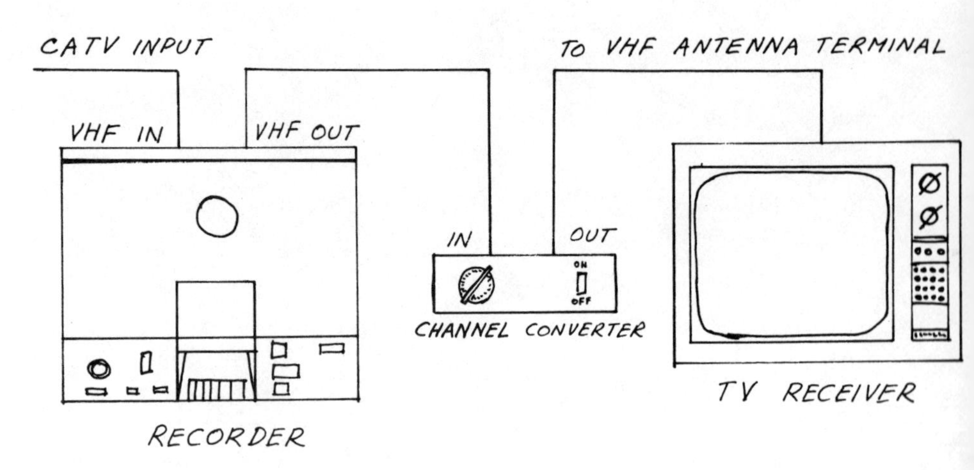

This cable TV system connection has the recorder first

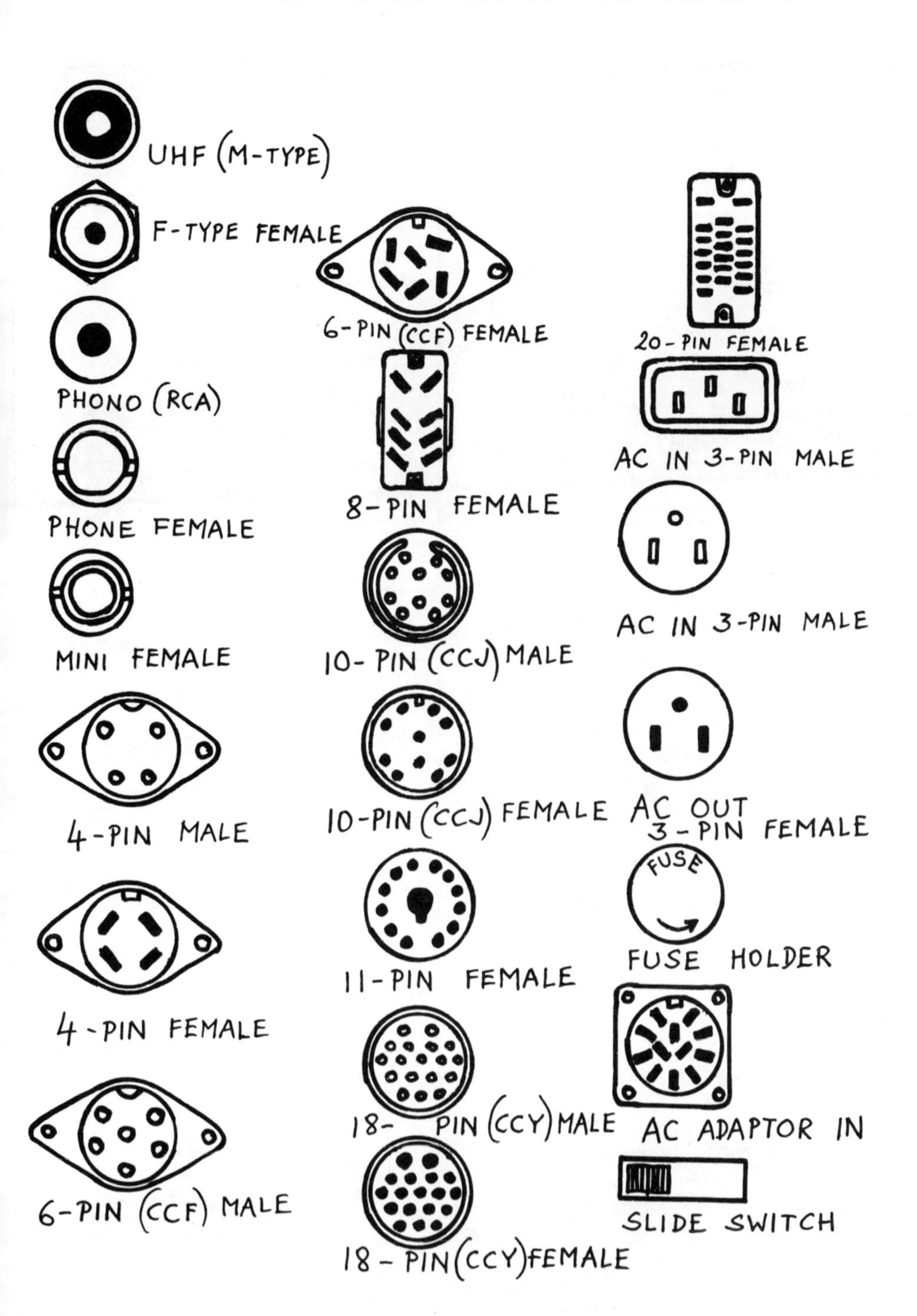

Typical connector symbols

INSERTION AND REMOVAL OF CASSETTES

INSERTION

1. Press the eject button to open the cassette compartment.

2. Insert the cassette. Be sure it is completely and correctly inserted.

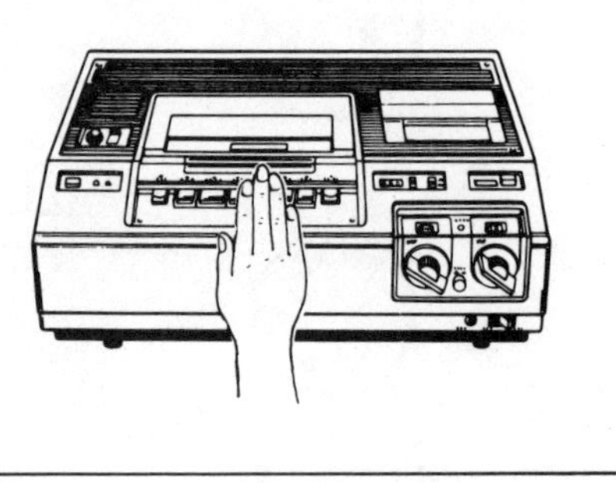

3. Press the cassette compartment gently downward all the way to close it.

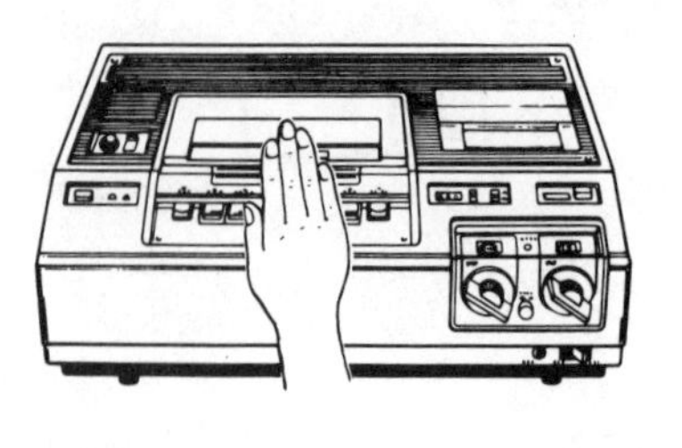

REMOVAL

1. Press the eject button to open the cassette compartment.

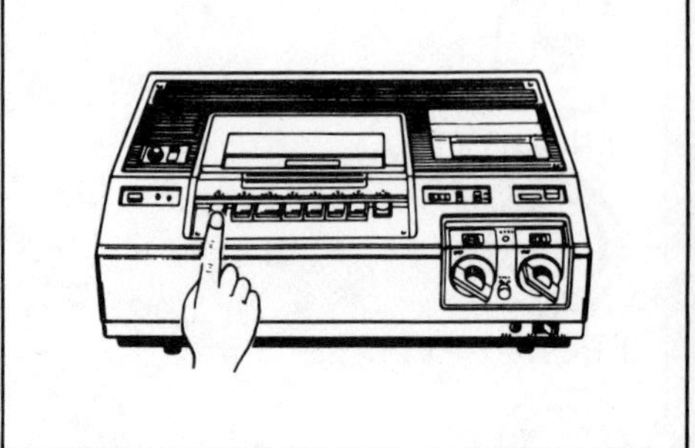

2. Remove the cassette.

3. Close the cassette compartment to prevent dust and dirt from entering.

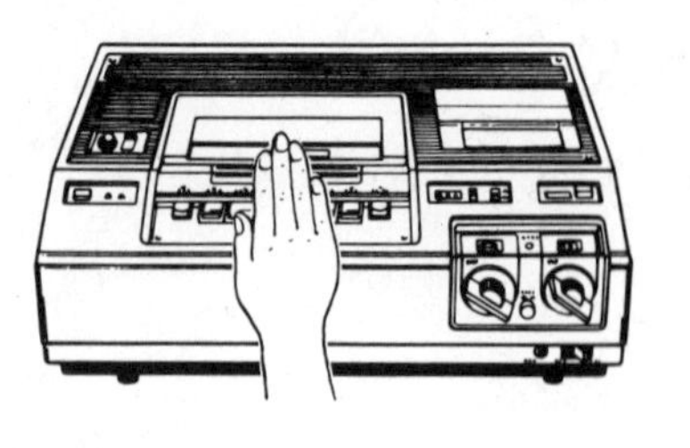

BASIC OPERATIONS

REWIND BUTTON

Press this button to rewind the tape.

PLAY BUTTON

Press this button to playback recorded tape.

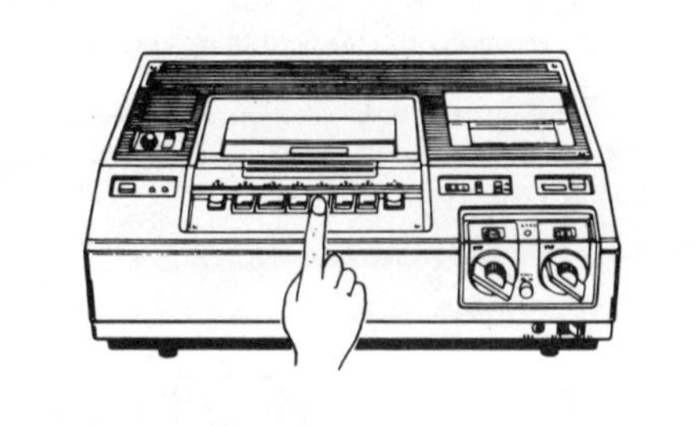

STOP BUTTON

Press this button to stop the tape.

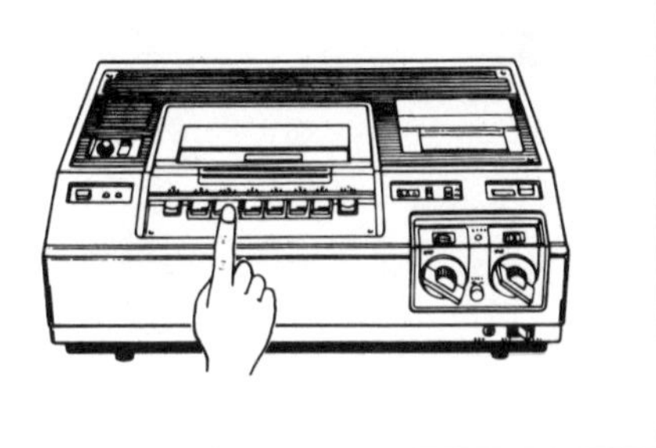

RECORD BUTTON

To record, press this button and the play button at the same time.

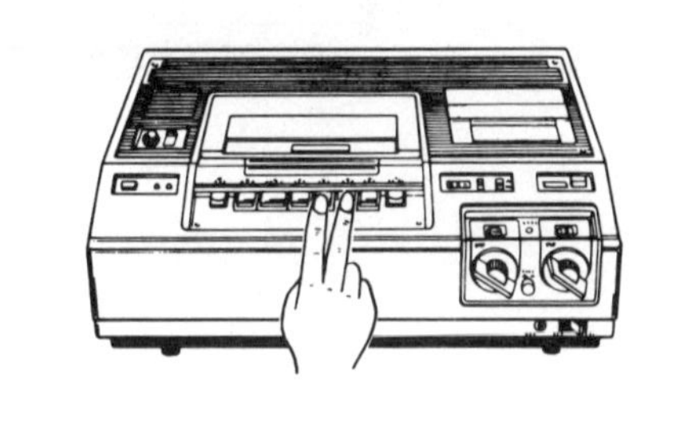

FAST FORWARD BUTTON

Press this button to move the tape forward rapidly.

PAUSE BUTTON

Press this button, during recording or playback, to instantly stop tape movement temporarily. Convenient to avoid recording of unwanted material.

THE RECORDING-TIME SELECTOR

The video cassettes used in this unit can be used to continuously record TV programs of as long as 4 hours. Set the recording-time selector to the desired position according to the program to be recorded.

Determine the length of the program beforehand, and then select the cassette to be used and the selector setting accordingly.

- This selector functions only during recording. During playback, the video deck automatically detects the speed at which the recording was made, thus making it unnecessary to reset the selector.
- The selector setting can be changed, if desired, while the recording is being made.
  If it has been reset during recording, the video image may momentarily disapper during playback. This is normal, and not a malfunction.

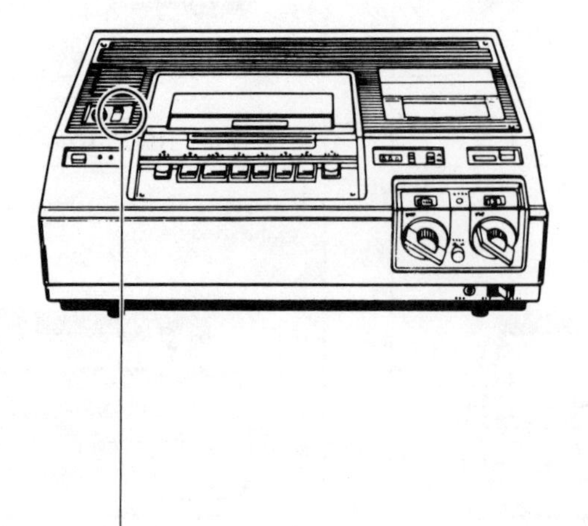

RECORDING-TIME SELECTOR	RECORDING TIME	
	T-120 Tape	T-60 Tape
SP	2 Hours	1 Hour
LP	4 Hours	2 Hours

RECORDING TV PROGRAMS

(1) Turn on the power to the TV, and set the VHF channel selector (of the TV) to the unused channel (3 or 4) in your area.

(2) Turn on the power to this unit. (The power-indicator lamp (red) will illuminate.)
Mack sure timer switch is in "OFF" position.

(3) Set the mode switch to the "VCR" position.

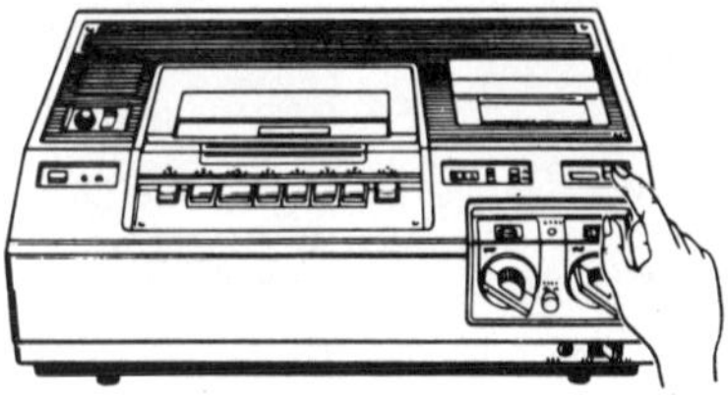

(4) Set the input selector to the "VCR TUNER" position.

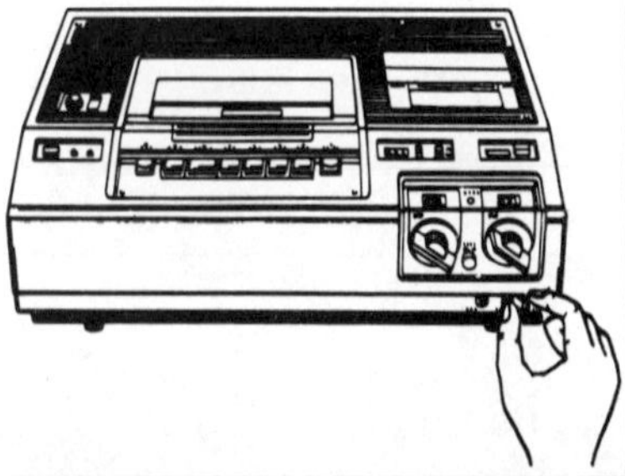

(5) Set the recording-time selector to the desired position according to the length of the program to be recorded and the type of cassette to be used.

(6) Insert the cassette.

(7) Turn the VHF (or UHF) channel selector of this unit to the channel which will broadcast the program to be recorded.

(8) Follow the fine-tuning procedures described below if the TV image is not clear.

(1) Set the AFC switch to the "OFF" position.

(2) Pushing the fine-tuning control inward, turn it until the best picture is obtained.

(3) After the adjustment is completed, once again set the AFC switch to the "ON" position.

If the video image is still not satisfactory, the fine-tuning of the TV itself should be adjusted.

(9) Press the record button and the play button at the same time.

NOTE: No sound will be recorded from the TV if a microphone is connected.

(10) The recording can be stopped by pushing the stop button. (The record button and play button will return to their original positions.)

(11) After the recording is finished, the tape can be rewound by pressing the rewind button.

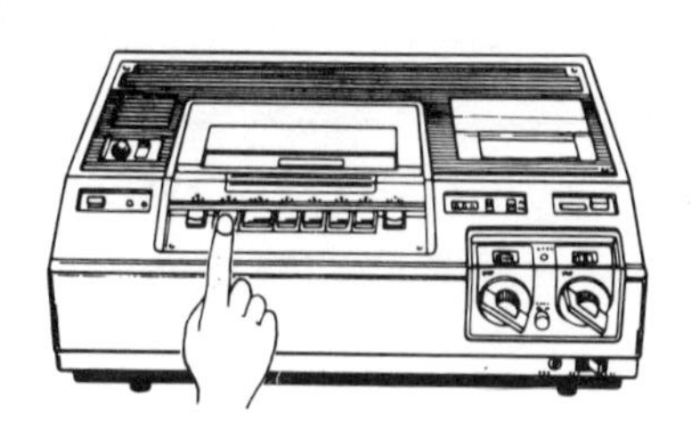

TO STOP TAPE MOVEMENT TEMPORARILY

The tape movement can be temporarily stopped during the recording by pushing the pause button. This is convenient, for example, to avoid recording unwanted material. Push pause button again to restart tape.

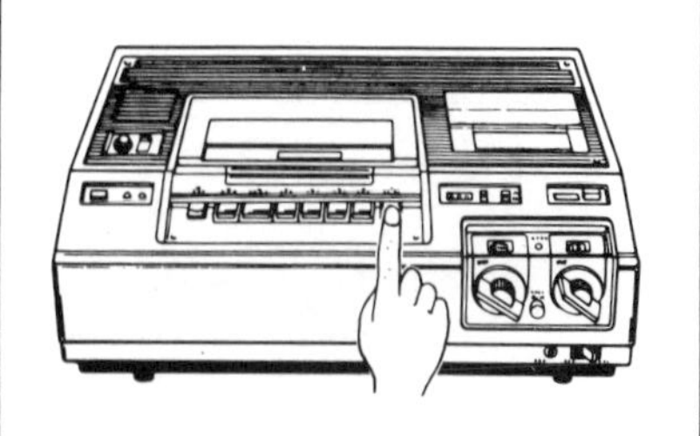

PLAYBACK

(1) Turn on the power to the TV, and set the VHF channel selector (of the TV) to the unused channel (3 or 4) in your area.	
(2) Turn on the power to this unit. (The power-indicator lamp (red) will illuminate.) Make sure timer switch is in "OFF" position.	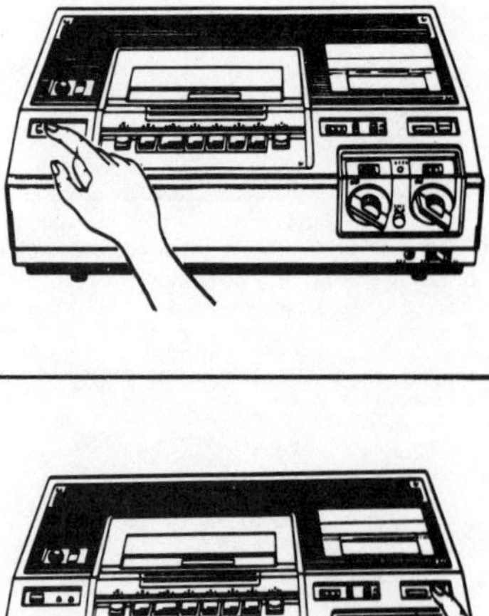
(3) Set the mode switch to the "VCR" position.	
(4) Insert the cassette.	

(5) By pressing the play button, the recorded material on the tape can be seen on the TV screen.

There is no need to set the recording-time selector, because the deck will automatically detect the recorded tape speed and playback at that speed.

(6) The playback can be stopped by pressing the stop button. (The play button will return to its original position.)

(7) Adjustment of the playback picture.

Playback image is partially bad. .... Turn the tracking control until the best picture is obtained.

The playback image is completely bad. .... The video heads need to cleaned or replaced. Contact your dealer.

RECORDING FROM ONE CHANNEL WHILE WATCHING ANOTHER

Although two programs which you want to see may be scheduled for broadcast at the same time, you can enjoy viewing one while using this unit to record the other for viewing at a later time.

OPERATION PROCEDURE

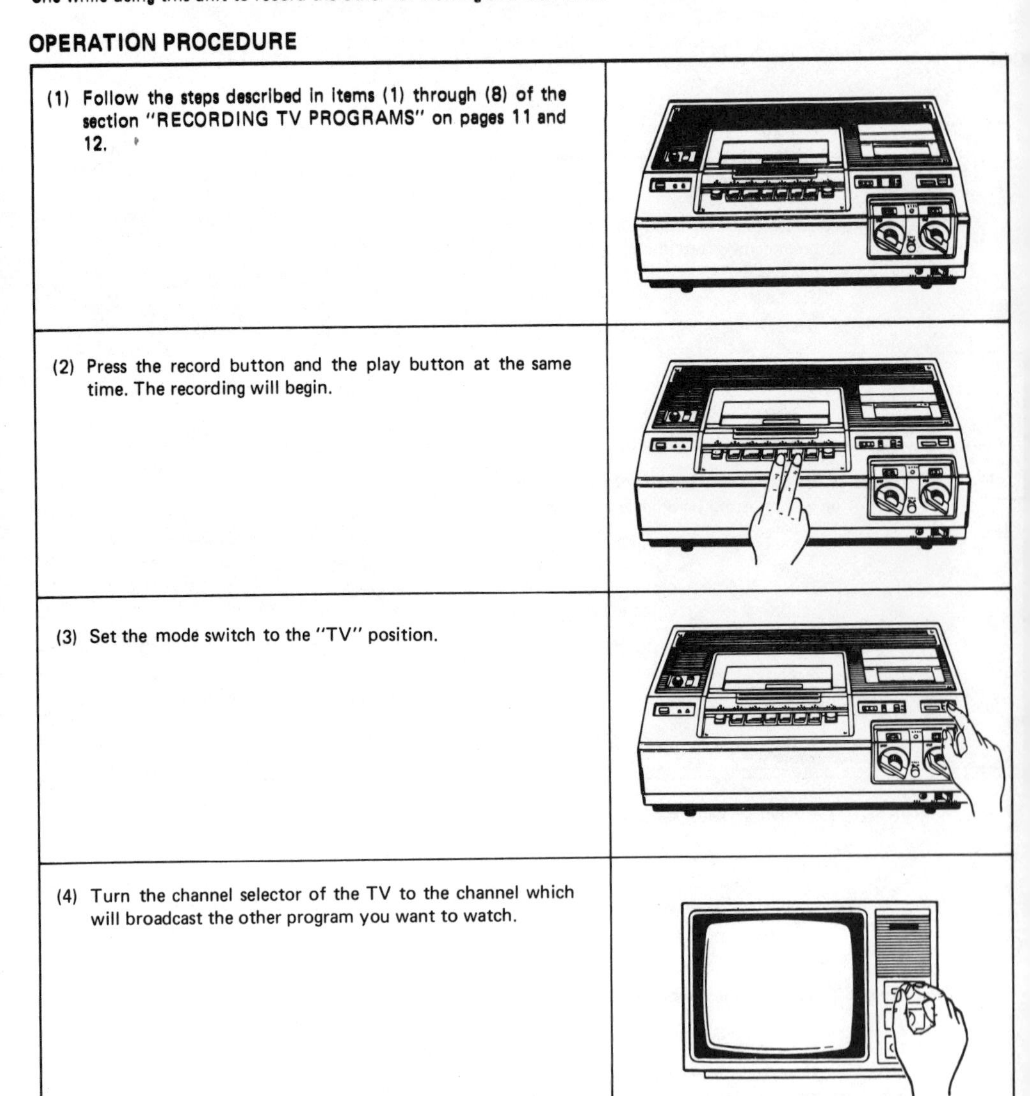

(1) Follow the steps described in items (1) through (8) of the section "RECORDING TV PROGRAMS" on pages 11 and 12.

(2) Press the record button and the play button at the same time. The recording will begin.

(3) Set the mode switch to the "TV" position.

(4) Turn the channel selector of the TV to the channel which will broadcast the other program you want to watch.

HOW TO THEN CHECK THE PROGRAM BEING RECORDED

(1) Set the mode switch to the "VCR" position.

(2) Set the channel selector of the TV to the open channel (3 or 4) being used for recording.

RECORDING FROM A VIDEO CAMERA

By connecting a video camera to this unit, you can easily enjoy producing your own TV programs, showing everyday life at home, your golf swing practice for later critique, a dance party, etc.

Refer to the operation instructions of the video camera for detailed information concerning connections and operation.

CONNECTIONS

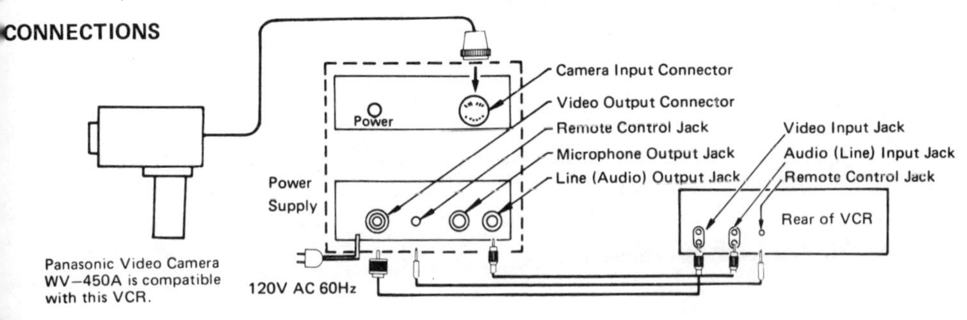

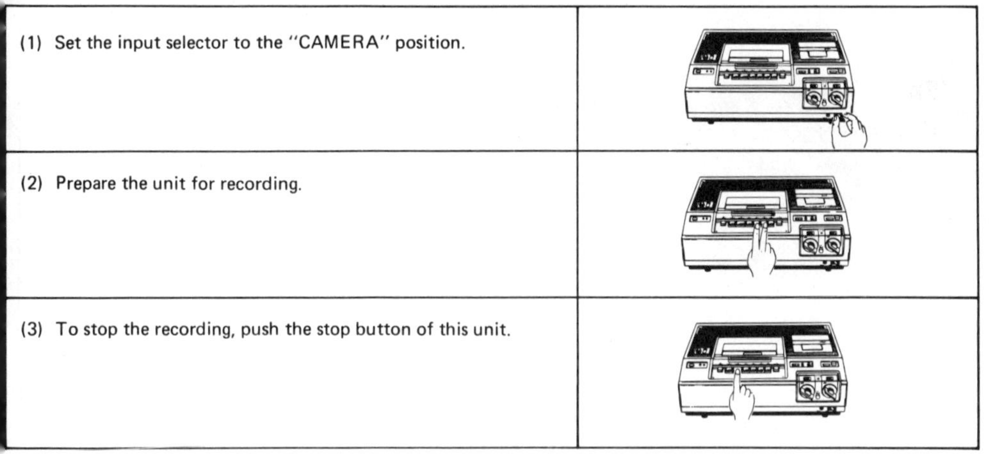

(1) Set the input selector to the "CAMERA" position.

(2) Prepare the unit for recording.

(3) To stop the recording, push the stop button of this unit.

If the video camera does not have a built-in microphone, connect a microphone to the unit in order to record audio at the same time.

If the video camera includes a remote-control unit, the recording can be started and stopped by connecting it to the remote-control connection terminal of the unit.

CAUTIONS

A. When the remote control switch or camera unit is connected to the VCR, the remote function will only activate the "Start/Stop" of the record or playback mode of the VCR unit.

B. The VCR will not playback unless the remote control switch or the camera unit is "on".

C. When the camera is connected to the VCR and the power of the camera is "off" the remote control of play back will not function. Disconnect the remote control cable from the VCR, then the VCR can be used normally.

NOTES: No recording can be made from both a microphone and a record player (or a tape recorder) at the same time. The sound from the microphone will take priority.
Be sure that the accidental erase-prevention tab (on the back of the cassette) is intact. If it is not, no audio dubbing can be made.

VIDEO TIMER PREPARATIONS

[1] To set the timer to the correct time.

(1) Set the timer switch ① to the "OFF" position.

(2) Set the time mode switch ② to the "TIME" position.

(3) Use the adjust switch ③ to set to the correct present time.

FF . . . . Time indication is advanced about 1 hour per second.

FWD . . Time indication is advanced about 2 minutes per second.

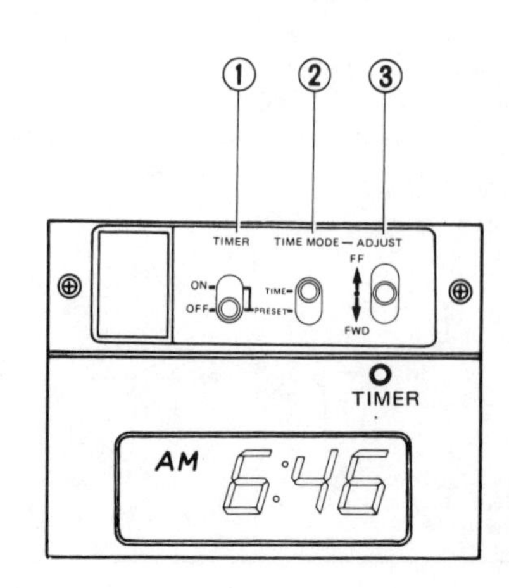

NOTES CONCERNING TIME SETTING

In order to assure proper operation of the timer, always advance the time by more than 24 hours, and then to the correct time when making the first setting after the power cord is connected or after a power failure.

[2] To prepare the timer for unattended recording.

(1) Set the timer switch to the "OFF" position.	TIMER ON— OFF—
(2) Set the time mode switch to the "PRESET" position.	TIME MODE TIME — PRESET —
(3) Use the FF and FWD settings of the adjust switch to set the time indication to the desired turn-on time. (Be sure the A.M. or P.M. setting is correct.)	ADJUST FF FWD
(4) Set the timer switch to the "ON" position. The timer pilot lamp will illuminate.	TIMER ON — OFF —

To again show the present time, set the time mode switch to the "TIME" position.	TIME MODE TIME — PRESET—

NOTE : If the AM or PM indicator flashes, the indicated time is incorrect, because there has been a power interruption, therefore the time should be reset to the correct time.

TIMER RECORDING

The built-in electronic timer can be used to begin unattended recordings even if you are away from home or asleep.

OPERATION PROCEDURE

[1] Refer to items (1) through (8) of the section "RECORDING TV PROGRAMS" on pages 11 and 12 to prepare the unit for recording.

[2] Prepare the timer for unattended recording.

(1) Set the timer switch to the "OFF" position.	TIMER ON – OFF –
(2) Set the time mode switch to the "PRESET" position.	TIME MODE TIME– PRESET–
(3) Set the timer to the desired turn-on time.	ADJUST FF FWD
(4) Set the timer switch to the "ON" position. (The timer pilot lamp will illuminate, and the power-indication lamp of this unit will go out, although the power switch will still be pushed.)	TIMER ON – OFF –
(5) Set the time mode switch to the "TIME" position.	TIME MODE TIME– PRESET–
(6) Press the record button and the play button at the same time. (The tape will not move however.)	
(7) Turn the TV set power off.	

By making the above settings, the recording will begin automatically at the selected time. The tape will stop automatically when it reaches its end, the record and play buttons will return to their original position, and the power will be cut off (although the power switch will remain pushed).

NOTES: If the power is interrupted during the recording, the recording will not continue after the power is once again restored.

After a timer recording, no other operations of the deck will function until the timer switch is set to the "OFF" position.

AUDIO DUBBING

Audio dubbing means to add new audio (voice or music), from a microphone, record player or tape recorder, to a recorded tape, thus erasing the original sound.

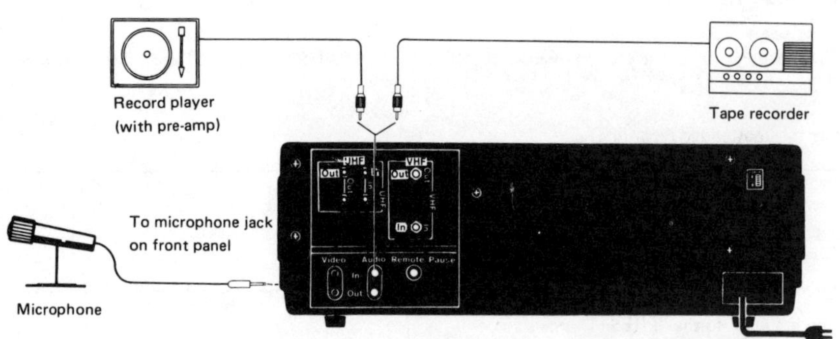

(1) Connect the desired audio dubbing source to the proper input connector. Set the input selector to the "CAMERA" position.	
(2) Prepare the unit for playback, referring to steps (1) through (5) of the section "PLAYBACK" on pages 14 and 15.	
(3) Watch the playback, and when it reaches the point where you want to make the audio dubbing, press the audio-dubbing button. The new audio is now being recorded.	
(4) To stop the audio dubbing, push the stop button of this unit. Disconnect the dubbing source.	

NOTES: No recording can be made from both a microphone and a record player (or a tape recorder) at the same time. The sound from the microphone will take priority.

Be sure that the accidental erase-prevention tab (on the back of the cassette) is intact. If it is not, no audio dubbing can be made.

OTHER FUNCTIONS

AUTOMATIC STOP

The tape will automatically stop when it reaches either end, and the operation button(s) will return to the original position.

MEMORY SWITCH

The memory switch can be used, during the rewind, to automatically stop the rewind when the tape counter reaches the pre-set "000" reading.
This can be used, for example, to automatically return the tape to a pre-selected position immediately after recording, or to repeatedly playback a portion of the tape.

■ **Operation Procedure**

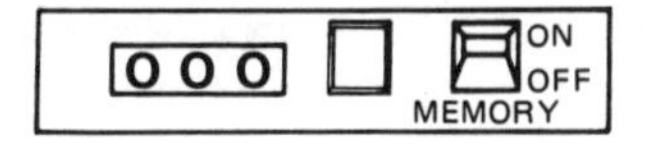

(1) Before (or during) playback or recording, set the tape counter to the "000" position at the beginning of the part you want to see after the rewinding.

(2) Set the memory switch to the "ON" position.

(3) Press the stop button when the recording or playback finishes.

(4) Press the rewind button to begin the rewinding, the tape will automatically stop at the pre-selected "000" position.

Actually, so that you won't miss the very first part of the program to be viewed, the tape will stop slightly prior to "000" at the "999" position.

CONCERNING MOISTURE ACCUMULATION

If moisture condensation accumulates inside this unit, the dew-indicator lamp will illuminate, and the unit will not function. This moisture accumulation is the same phenomena observed during the winter when moisture forms on windows of a heated room. It may occur in this unit under the following conditions:

- In a room where there is excessive steam or humidity.
- When the unit is suddenly brought from a cold place into a warm room.

If this dew-indicator lamp illuminates during operation, the unit will automatically switch itself to the "STOP" mode. The unit will not function until the lamp goes out. At the very longest, the moisture will disappear in a few hours.

PAUSE

The pause button of this unit can be used, during recording or playback, to temporarily stop tape movement. This pause operation can also be accomplished from a distance away from the unit by connecting a remote control unit (supplied) to the remote control connection terminal of this unit. (Fast forward and rewind cannot be operated by remote control.)

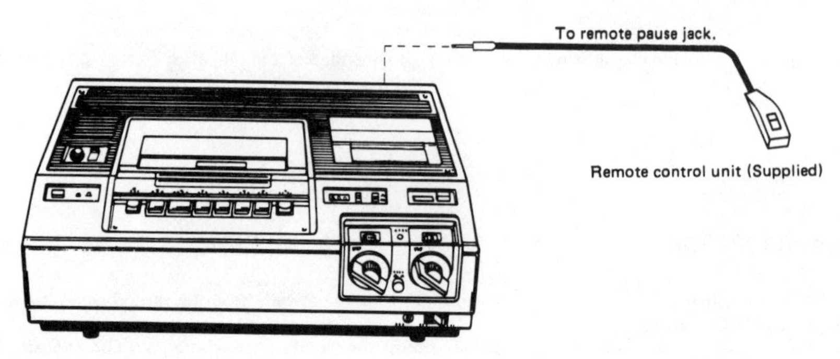

■ **Pause during playback**

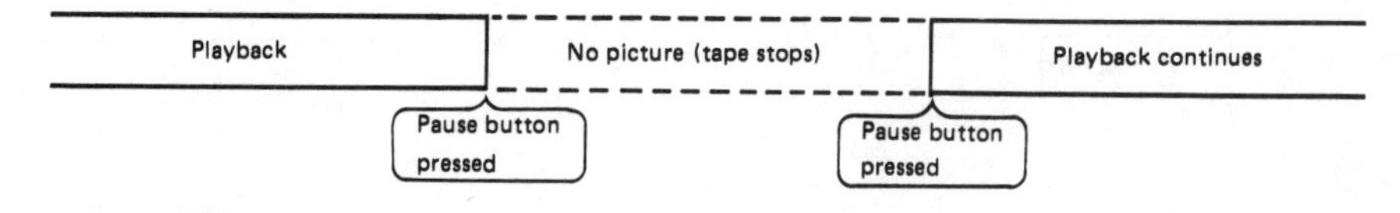

■ **Pause to change from playback to recording**
(or to add a recording to already recorded tape, etc.)

Playback | Recording preparations (tape stopped) | Recording

Pause button pressed | Record & Play buttons pressed | Pause button pressed

■ **Pause during recording**
(to omit unwanted material, or to connect video camera to add recording)

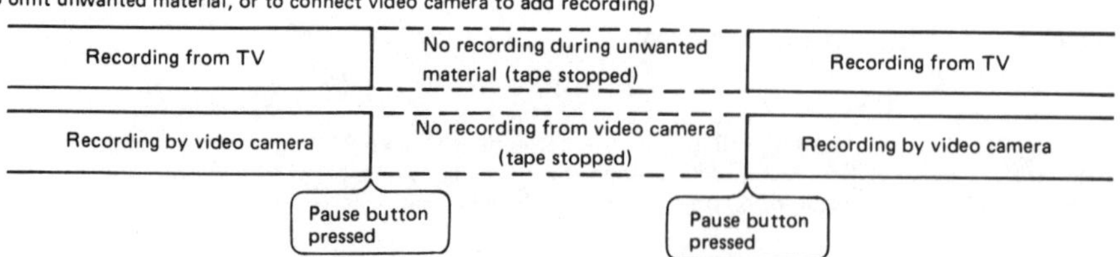

NOTE: To prevent tape damage, do not leave the unit in the pause mode for more than 5 minutes.

BEFORE REQUESTING SERVICE

CHECK THE FOLLOWING POINTS ONCE AGAIN.

CONDITION	MAIN CAUSE AND REMEDY
(1) No power.	• Power cord may be disconnected. • If timer is being used, timer switch may be set to "ON" position. 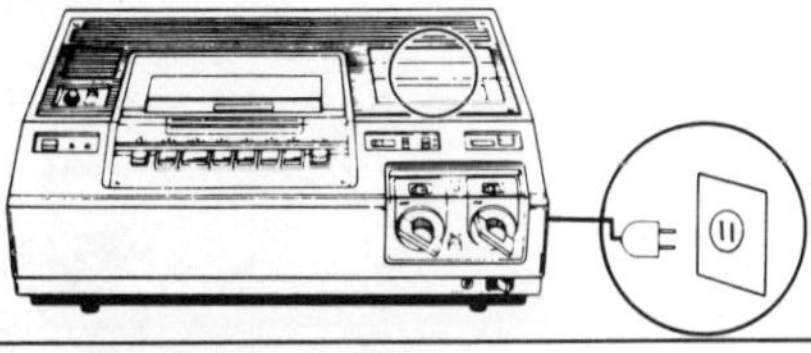
(2) Video cassette cannot be inserted.	• Inserting of cassette is not correct. 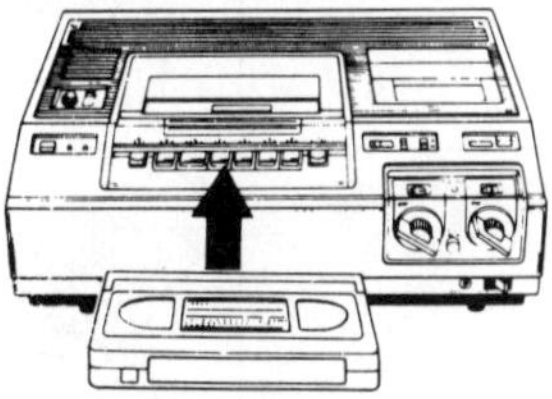
(3) No operation when button(s) pressed.	• Power switch may be off. • Is dew-indicator lamp (yellow) on? • Check for no power (1) above. 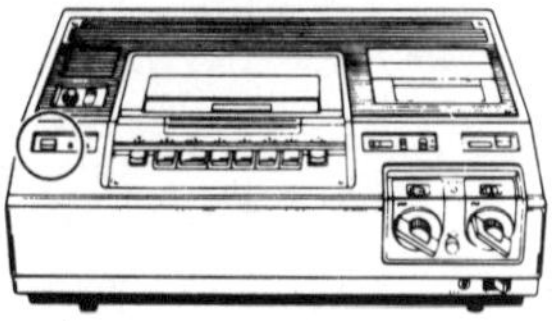
(4) Record button cannot be pressed.	• Is accidental erase-prevention tab broken out? 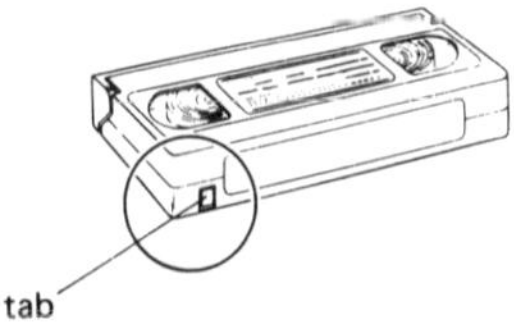 tab
(5) If tape can't be rewound or fast forwarded.	• Is the tape already completely rewound (or fast forwarded 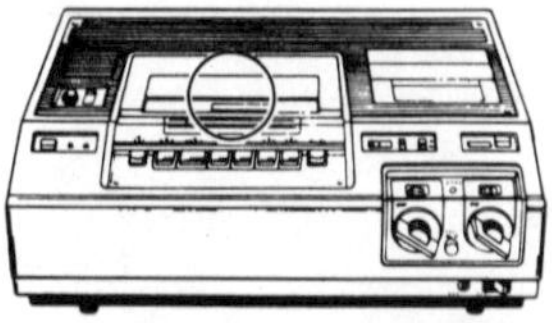) ?

CONDITION	MAIN CAUSE AND REMEDY
(6) If TV programs can't be recorded.	• Is the Input Selector set to the "VCR TUNER" position? • Is the antenna connected correctly? Has the tab been removed? 
(7) If there is no playback picture.	• Is the mode switch set to the "VCR" position? • Is the VHF channel setting of the TV the same as that of the RF converter (channel 3 or channel 4)? Channel 3 or Channel 4  
(8) If the playback image is noisy or contains streaks.	• Turn the tracking control. 
(9) If there is no color in the playback image.	• Was the program broadcast in black and white? • Was the VHF Channel Selector VHF Fine Tuner of this unit adjusted before the recording was made? • Is the fine-tuning adjustment of the TV correct? 
(10) Does the top of your playback picture wave back and forth excessively?	• Locate the horizontal hold control on your TV set. Turn it slowly to correct for the wavy picture.  Adjust horizontal hold

Refer servicing to qualified service personnel.

the cassette. The tape comes out of the cassette through the exit guide. The tape passes the sensing head mounted on the tension arm assembly. When all the tape has wound onto the take-up, that is, attached to the end of the video tape and which appears after the normal tape passes the sensing head, the sensing head senses the trailer tape and the forward or fast-forward mode is terminated.

The tape passes the tension arm assembly, which prevents tape slack between the exit guide and the supply tension regulator. It also acts to keep a constant clearance between the tape and sensing head.

There are two tape guide bosses (upper and lower) on the tension arm assembly to regulate the height of the tape wound by the supply reel in the rewind mode.

The tape, having passed the tension arm assembly, passes the supply tension regulator arm assembly. The supply tension regulator arm assembly changes the forward direction of the tape toward the drum. At the same time, it senses tape tension in the forward operation and serves to control the braking pressure to the supply reel table assembly through the brake band assembly. This keeps a uniform holdback tension on the tape. The right angle of the pin on the supply tension regulator arm assembly assures proper tracking at the entrance of the drum. The tape then passes rotary tape guide *A*, the full-width erase head, and tape guide *B*.

The full-width erase head erases the tape during the record mode. Rotary tape guide *A* and tape guide *B* regulate the tape on its upper and lower edges in order to afford proper tracking from the first contact point of the tape to the drum to the middle point 90 degrees around the drum. Tape guide *B* serves to provide a constant overlap.

The tape runs approximately 180 degrees around the circumference of the drum where the rotary video heads and makes contact with it during record or playback. The tape path is designed to be parallel with the reel tables and at the entrance side of the drum and to be parallel with the loading board at its exit. The tape retaining springs, at the middle section of the drum, press down on the tape running along the drum in order to maintain tracking at the middle section.

The tape, having passed the drum, passes tape guide *C*, and the audio and erase heads and tape guide *B*. Tape guides *C* and *B* regulate the tape position from the top in order to maintain tracking from the middle section of the drum to the exit point of the tape.

Tape guide *C* serves to keep a constant overlap. The audio/control/erase head assembly performs erasure, record, and playback of the audio signal, and the record and playback of the control signal. The tape is adjusted with respect to the upper flanges of tape guides *C* and *P*. It is so designed that no height adjustment of the audio head in the assembly is necessary if the tape runs along the upper flanges of tape guides *C* and *P*.

The tape is squeezed between the capstan housing assembly and the pinch roller assembly. The capstan housing assembly rotates at a constant speed, advancing the tape at a fixed rate. The angle of the capstan assembly against the tape is very precisely set to degrees. The pinch roller is pressed against the tape at an angle of 90 degrees with respect to the forward direction of the tape with automatic alignment in order to make the tape run in a stable manner. The tape, fed at a constant speed, passes the guide roller assembly and the forward direction of the tape turns about 180 degrees in order to reverse the running direction of the tape.

The guide roller assembly is designed to turn freely so that the friction between the guide roller assembly and the tape is minimized for reduced tape wear.

The tape, having passed the guide roller assembly, passes the tape-slack sensing lever assembly which senses any tape slack caused by reduced tape tension in the forward mode. The slack sensing assembly terminates the forward mode and sets up the stop mode.

The tape passes the fixed tape guide. The space between the upper and lower flanges on the fixed tape guide is narrower than the spaces of other flanges because it is designed so that the tape height is in the fast forward, or rewind mode, or at the completion of the threading operation. Thus the tape runs smoothly at the beginning of the forward mode. The tape passes the loading arm assembly, the forward direction of the tape turns right about 60 degrees, and the tape is positioned at the same height as that of the take-up reel table assembly. The tape then passes the take-up sensing head. When all the tape is wound onto the supply reel in the rewind mode the leader tape appears at the beginning of the video tape.

The sensing head senses the leader tape and the rewind mode terminates. The tape, having passed the sending head, enters into the cassette, passes the exit guide, the tape slack prevention plate, and the guide, and finally is wound by the take-up reel table assembly. As described above the tape advances from the supply reel tables assembly to the take-up reel table assembly. Since the take-up reel table rotates in the forward, fast forward, or rewind mode, and the counter belt moves with it, the counter also functions.

General Information

1. The still, Playback, and Pause Buttons in recording put slightly more burden on the tape than normal playback and recording.

2. Frequent switching from play to stop and from rewind (or fast forward) to stop generate irregular windings on the reel. This may cause a decreased rewind or fast-forward-type speed, abnormal skew, or increased jitter.

3. It takes some time to become accustomed to the timer setting for automatic timer recording. Until then, set the timer to a time a little before the desired recording time.

4. The tuner should be used only for recording TV programs on video cassettes.

5. The Museum of Broadcasting (located at 1 East 53 Street in New York City) has an impressive collection of old network TV and radio shows; anyone may view them for a one-dollar donation. Members can reserve shows or use of a console, and can spend the whole day there. Membership is $30 per year, $20 for students or anyone who lives more than 50 miles away.

6. U-Matic (¾″) still remains the best format for storing

masters of programs you wish to keep and from which to make copies. The audio frequency response is noticeably better (than Betamax or VHS) on U-Matic, especially for high-quality musical programs; and stereo simulcasts are possible only on U-Matic.

7. *Question:* Is the Sony KC 60 tape the same as the L-500 except for a more attractive box?

Answer: No, the L-500 has a special lubricant to keep down "chatter."

8. *Question:* I bought a Betamax in the United States. Can I play it back on my set in England?

Answer: No. English television has 625 scanline; United States sets have 525 scanline.

9. Allied Artists, the film distributor, has made available on video cassette its entire library of 50 films. Allied Artist movies include *Cabaret, Papillon, The Man Who Would Be King, The Story of O,* and the series of Charlie Chan flicks.

How to Make the Perfect Edit to Eliminate Commercials or Station Breaks

1. You're recording your program.
2. Program fades out, commercial begins. Let the tape roll for a few seconds as commercial progresses.
3. Hit stop, then rewind for 3 to 5 digits on the counter. Hit stop again.
4. Push the play button and watch until your program fades out. Then hit *pause,* not stop. Lock the pause lever back. You are now up to the point where the program has ended.
5. With the pause button engaged, push the record button.
6. As soon as you see your program is about to continue, release the pause button.
7. Nine out of ten times you will wind up with a perfect edit.

5 • *Maintenance and Troubleshooting*

General Care and Simple Maintenance of the VTR

In this chapter we will cover those things an educated consumer can do to look after his or her video cassette recorder, to keep it running in good order.

Test Tape

First make yourself a test tape. A test tape consists of a recording, made by the operator and therefore known to contain good video and good audio tracks. Thus you have a tape known to be good, which can be played back through any system suspected of having problems. This device will remove any suspicion that the tape may be defective and not the machine.

A test tape should consist of color bars: if not available, any stable picture will do. For the best audio track, music should be used; even piano music or simple tones will work.

Alignment Tape

Alignment tape is used for the final adjustment of the electronics. It should never be used to check out a suspected or defective machine; such a check could easily ruin your alignment tape. The alignment tape should be kept in a safe place and marked with a bright-colored tape.

Cameras

Cameras must conform to American TV (EIA) standards.

Connections

The microphone connector has priority over the audio line in connector. Any input to the audio line in connector is automatically disconnected whenever a microphone is inserted into the microphone connector.

The audio line in connector has priority over the VHF in connector. TV audio input to the recorder is automatically disconnected when a external audio source is connected to the audio line in connector.

The video in connector has priority over the VHF in connector. TV video input to the recorder is automatically disconnected when an external video source is connected to the video in connector.

Camera Recording

In camera recording, connect a microphone to the microphone connector, or connect a tape recorder or a record player to the audio line in connector. If no signal is connected to the microphone and audio line in connectors, TV sound is recorded.

Cautions

Before making a TV recording, check that the cables from the video in, microphone, and audio line in connectors are disconnected.

The Work Area

The work area where the video cassette is used should be as clean as possible. Discourage any persons from eating, smoking, or drinking, or having dirty or greasy hands, around a VCR. Grease from the fingers can adhere to the tape and the machine parts, gathering dust and dirt. Once dust or dirt gets into the tape, it will make contact with the video head during play or record and transfer the dirt onto the video head.

In particular, observe the following precautions:

1. Do not allow excess moisture to enter the machine or tapes.

2. Do not cover the air holes in the machines by putting things over them.

3. Always handle the machine very gently. Do not bang it around.

4. Do not drop cables and plugs onto the floor. This can bend and shatter them.

5. Do not pass cables through doorways which are allowed to fully close. This invites broken cables and damaged plugs. For the same reason, do not step on them or allow heavy equipment to roll over them.

6. Never use damaged tape in a machine. It must be removed and discarded.

7. Never operate the machine right after having transported it from a cold location directly to a warm location.

8. Operate the machine on 120V AC, 60 Hz.

9. Allow adequate air circulation to prevent internal heat build-up.

10. Unplug the machine from the wall outlet when it is not to be used for an extended period of time.

11. To disconnect the cord, pull it out by the plug. Never pull it by the cord.

12. The machine is designed for operation in a horizontal position.

13. Save the carton and associated packing material. They will be useful should you have to transport or ship the machine.

14. Avoid places where the unit will not be level, where it will not be stable, and where it might be subjected to vibration.

15. Do not transport the VTR with a video cassette in place.

16. Do not turn on the set with dust cover in place.

17. Allow adequate air circulation to prevent internal build-up. Do not place the VTR on surfaces such as rugs, blankets, and the like, or near such materials as curtains or draperies.

18. Do not install the set near heat sources such as radiators or air ducts, or in a place subject to direct sunlight, excessive dust, mechanical vibration, or shock.

19. Keep the set and the video cassette away from strong magnetic fields.

20. Avoid subjecting the machine to unnecessary shock or impact.

21. Should any liquid or solid object fall into the cabinet, unplug the set and have it checked by qualified personnel before operating it any further.

Cleaning the Machine

Cleaning is very important to any VTR. Two main things arise in cleaning. The first is oxide shed from the tape. The second is environmental dust, dirt, and grease. The build-up of oxide shed from the tape occurs over a period of time and must be cleaned off. If not, it will gradually affect the tape path, and the interchange of the tapes will be ruined. In more serious cases the heads can completely clog, and no recording or playback will be possible until they are cleaned.

Video Head Cleaning

The most important and expensive parts of the VTR are the video heads. These delicate parts must be treated with the utmost care and respect for cleanliness. To clean your video head, use a special head-cleaning pad (available through Sony). Cotton-tipped household swabs should not be used to clean the video heads as their fibers can catch on the heads and pull on them, causing them to chip or break. The special head cleaner pad should be soaked in a cleaning fluid or pressurized aerosol (feron) and then gently and firmly rubbed sideways across the head. Never rub it up and down; this may cause the video heads to break. The whole drum should be cleaned sideways. If you are not up to this, use a video head cleaning cassette. This special video head cleaning cassette contains a mildly abrasive tape which will remove the excess oxide and dirt from all the places the tape touches.

Never spray the feron directly onto a hot rotating head; it can cause damage. Instead, spray it onto a lint-free cloth. Xylene is good for cleaning metal components; however, do not let it touch any rubber parts, such as the pinch roller. Denatured alcohol is the best solvent for rubber parts. But the denatured alcohol available in drug stores, such as rubbing alcohol, contains oils, and should not be used for machine cleaning. Use only denatured alcohol obtained from a video store.

NOTE:

Never try to clean the rotary video heads while the motor is running.

Video Head Cleaning Cassette

Follow these steps in sequence:

1. Insert the cleaning cassette into the cassette compartment as you would a video cassette.
2. Reset the tape counter to "000."
3. Press the FWD or play button and watch the tape counter.
4. Press the stop button when the counter indicates "010" (about 30 seconds running time).
5. Remove the cleaning cassette. Do not rewind the cleaning cassette at the end of each use. The cleaning cassette may be rewound and reused several times. However, after prolonged use, the cassette will lose its effectiveness and will require replacement.

CAUTION

Excess use will wear out the heads and the guides faster, so its use should be severely limited to about 10 times.

Degaussing Video Heads and Other Parts

If the video head is magnetized, the S/N ratio deteriorates and slant beat and noise appear on the picture. The video head and other parts must be demagnetized. To do this, bring the tip of the demagnetizer as close as possible to the head tip without actually contacting the head tip. Withdraw the demagnetizer very slowly and turn off the power of the demagnetizer when it is at least 7 feet away from the deck.

Cleaning the video heads *(Photo by Vincent Bellotti)*

Tape Disentangling

Occasionally, the tape will become entangled inside of the machine. The utmost care must be used so that the video heads are not damaged.

First, depower the machine; the tape is then free from the mechanism and is held by hand so that it forms a loop with its ends disappearing back into the cassette. Then press the stop button, and power is reapplied; the unthreading process will complete itself and the tape will retract into the cassette. Eject the cassette and hold the cassette in your hand. Look for a release level in the cassette that will open the protective flap. Examine the tape.

If the tape is badly damaged, cut the damaged portion out and splice the ends together. Make certain you put the splicing tape on the side away from the video heads. Place the video cassette back into the recorder, run the tape to the end by using the forward operation, and then rewind the entire tape on the supply reel. This eliminates any irregular wind in the cassette.

Tape Tension

When tension is applied to a rubber band it stretches; this applies to video tape as well. Thus video tape tension changes will cause a tone or pitch variation and wow flutter problems. Tension variations will cause the tape to skew differently across the heads, causing tracking problems. This problem is minimized when a tape is played back on the machine on which it was recorded. Since the tension will change in the same manner during playback the skewing error will cancel itself out. Playing a tape recorded with constant torque will cause problems when played back on a newer constant-tension machine (or vice versa). Tension differences will cause program time errors. Tape stretched longer by higher tension will take longer to play back. Conversely, tape stretched longer during record will not take as long to play back when used at a machine's proper tension.

A modern method of checking tension is by simply removing the top of the machine cover and checking tensions dynamically while the machine is operating. The device to use is called a tentelometer, manufactured by the Tentel Corporation

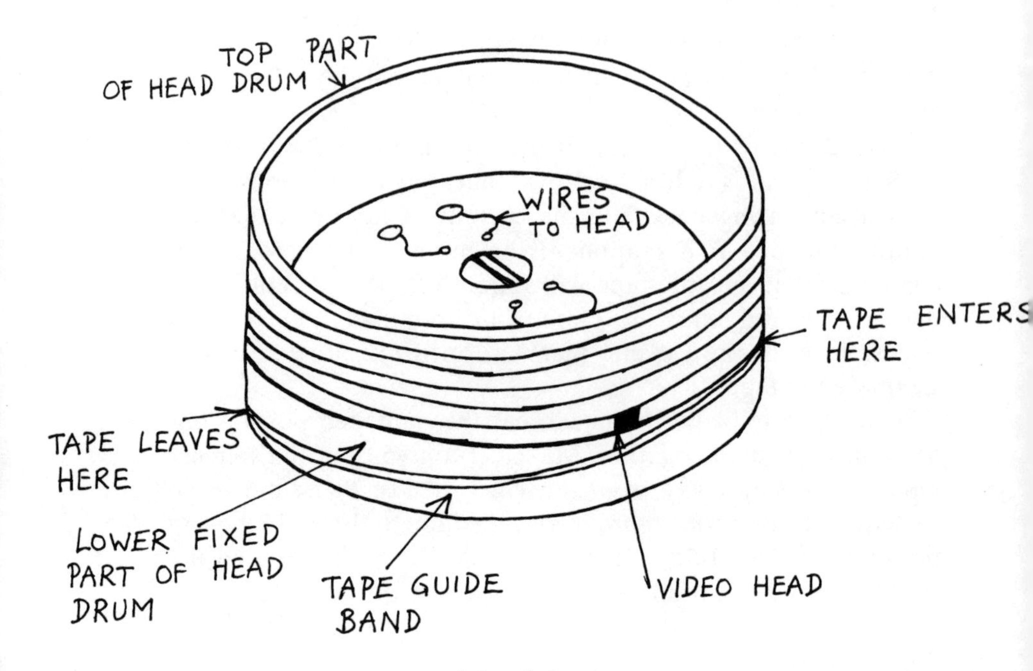

A head drum

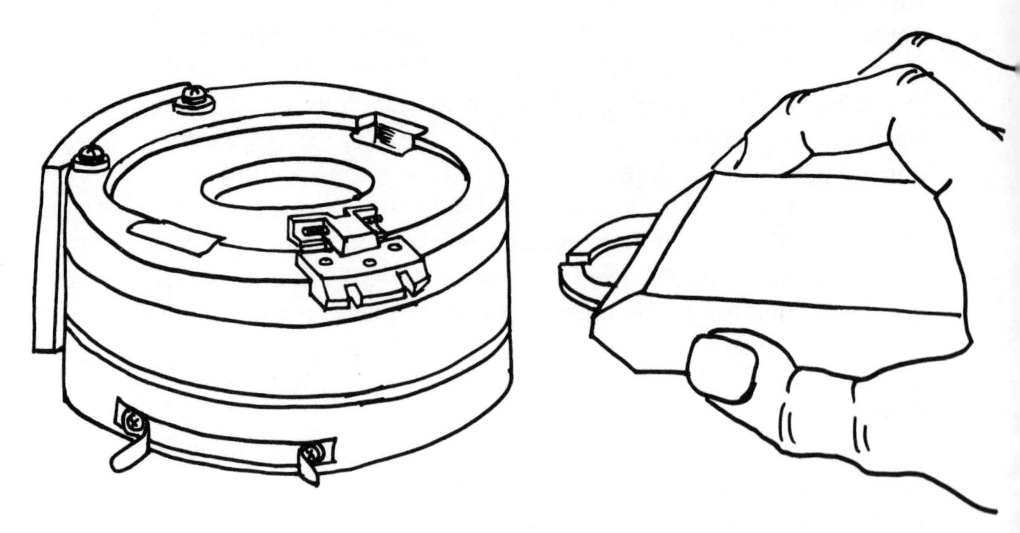

Degaussing of the video heads

(Campbell, California). With the tentelometer, all machines with such a facility can be set to the same tension, eliminating tape stretches.

Belts

When belts are new, they provide a good positive strive; but like everything else with age, they become brittle and tend to stretch. As belts wear, the servos can adjust to correct for the transport problems, but there is a point at which the belts are so worn that the servos can no longer adequately adjust themselves. This problem can be solved by changing the belts.

To replace a worn belt, see your service manual for the position of the major belts. Belt changing is very easy, but extreme care must be taken to avoid stretching the belts while placing them over the pulley wheels.

Signal-to-Noise Ratio

This is a measurement of the amount of "snow" or noise which can be seen in a picture. An S/N ratio of 45 dB means that no noise or snow can be seen by the human eye. An S/N ratio of 35 dB still allows an excellent picture, with traces of noise noticeable to a trained eye. An S/N ratio of 30 dB means the noise is becoming noticeable. An S/N ratio of 20 dB means the noise is easily noticeable.

If an S/N ratio gets as bad as 10 dB, then the picture is not viewable.

The common problem of noise is due to inadequate lighting in the studio or inadequate adjustment of cameras or other equipment at the time of recording. Noise is also caused by copying: passing the signal through equipment and transmission over long lines or over the air.

Noise is a high-frequency problem, and it can be moderated to some degree by lowering the high-frequency response of the system, but this lowers the overall fine detail of the picture. Noise is more easily tolerated in a color picture than in a monochrome (B/W) picture because the color hides much of the fine detail of a picture.

A tentelometer in use

Schedule for Periodic Maintenance

	Operating Hours								
Name of Part	*500*	*1000*	*1500*	*2000*	*2500*	*3000*	*3500*	*4000*	*5000*
Rotary drum assembly	C	C,R	C	C,R	C	C,R	C	C,R	C,R
Capstan assembly		C,L		C,L		C,L	C,R,L		C,R
Pinch roller	C	C		C		C,R		C	C
Drum belt		R		R		R		R	R
Capstan belt				R				R	
Forward belt				R				R	
Fast forward belt				R				R	
Counter belt				R				R	
Supply reel table assembly				C,L				C,L	
Take-up reel table assembly				C,L				C,L	
Forward idler assembly				R				R	
Fast forward assembly				C,R,L				C,R,L	
Brake band assembly				R				R	
Supply and take-up assembly				R				R	
Threading brake assembly				R				R	
AC motor						R			
Tape guides	C	C	C	C	C	C	C	C	C
Capstan motor						R			

Note: This chart lists the parts of the machine which should receive periodic maintenance, as well as the recommended intervals and type of maintenance required. C–Cleaning L–Lubrication R–Replacement

Troubleshooting for Failures in Playback or Recording

This section covers the common faults and problems likely to occur with your video cassette machine. The first thing to do is remove the top of the machine and explore the inside of the VTR. Now you can get a good idea of what is going on. Once the top of the machine has been removed, the first things to check for are:

1. Faulty key switches and buttons.
2. Relays and solenoids not working.
3. Linkage not moving.
4. The end of tape lamp not illuminated.
5. Tape tension loose.
6. Wiring problems (shorted or pinched wires).
7. Printed circuit board problems.
8. Burning odor (look for smoke or burned components).

No picture appears

1. Monitor TV is defective.
2. Monitor TV connection cable is open.
3. Input select switch of monitor TV is not in VTR.
4. RF selector to wrong TV channel.
5. RF Converter is defective.
6. Wiring to connector panel is open.
7. Wiring to converter is open.
8. Video head is dirty.
9. Video head or rotary transformer is defective.
10. Nothing has been recorded on tape.

Picture drifts or collapses

1. Synchronization of TV receiver is maladjusted.
2. Tracking is maladjusted.
3. Belt has partly come off motor.
4. Wrong power-line frequency.
5. Control head is dirty.
6. Tape transport mechanism is faulty.
7. Control head is defective.
8. Wiring to drum pulse head is open.

9. Drum pulse head is defective.
10. Electromagnetic brake is defective.

Picture is partly blanked

1. Tape is improperly in contact with video head at drum inlet or outlet.
2. Absolute height of video head is abnormal.
3. Switching phase adjustment.

Switching point has large bend (skew)

1. Back tension is abnormal (strong back tension causes left bend, weak back tension causes right bend).
2. Tape path is dirty.
3. Cassette insertion is abnormal.
4. Tension brake is maladjusted.
5. Tape has elongated.

Picture has no color

1. Color-mono switch (check).
2. May not be a color tape.
3. Heads dirty.
4. Tracking off.
5. Check color controls on TV.
6. Tension wrong.

No color recorded

1. Input signal not in color.
2. Defective record amplifier.

No sound is reproduced

1. Recorded channel is different from reproduced channel.
2. Audio select lever is in wrong position.
3. Audio head is in abnormal contact with tape.
4. Audio head is dirty.
5. Recording switch is defective.
6. Audio head is defective.
7. Wiring to RV converter is open.
8. RF converter is faulty.

Audio Problems

A. Hum in the audio:
   1. Bad ground or broken shield cables.
   2. Control track shield has broken.
B. Buzz in the audio:
   1. Antenna is misconnected.
   2. TV tuning is off.
   3. Video is overmodulated on the tape.
   4. The Automatic Gain Control on the tuner is not working.
C. Audio distortion:
   1. Bad audio on the tape.
   2. Automatic gain control (AGC) not working.
D. Noisy playback:
   1. Record level was too low on original tape.
   2. Dirty heads.
E. Low-level playback:
   1. Playback amplifier gain too low.
F. Overloaded audio:
   1. Playback amplifier gain too high.
   2. AGC may not be working.

Tuner section failure

A. No VHF/UHF image and sound:
   1. Input selector switch on connector panel is in TV position.
   2. Wires and connectors for coupling tuner section to main section are open.
   3. Antenna cable is shorted or input terminal section has poor contact.
B. Stripes appearing on screen:
   1. Tuner local is deviated.
   2. Antenna select switch operator is faulty.
   3. RF converter is faulty.
   4. Distributor contained in tuner section is defective or has faulty connection.

Time section failure

A. Timer lamp not lighted:
   1. Neon lamp is defective.
   2. Coordination of push switch and microswitch is faulty.
   3. Microswitch of timer is faulty.
   4. Power switch of VTR is not turned on.

B. No timer operation:
   1. Microswitch of timer is defective.
   2. Timer is defective.
   3. Coordination of push switch and microswitch is faulty.

C. No timer recording:
   1. Record lock microswitch operation is faulty.
   2. Starters on tuner video circuit go and are defective.
   3. Nine-pin connector for coupling tuner section with recorder section has poor contact.

Faulty Control System: Some Common Causes of Problems

Fuse blown when play switch is pressed

1. Filter for DC motor is shorted.
2. DC motor is defective.
3. Wiring to DC motor is in contact with chassis.
4. Diode of main dolenoid is shorted or its polarity is reversed.
5. Main solenoid wiring is shorted.

When eject lever is pulled during stop, cassette does not come up

A. Eject lock is not released:
   1. Loading arm is not reset.
   2. Main solenoid is not reset.
   3. Eject lock mechanism is faulty.

B. Eject lock lever is released:
   1. Cassette pop-up mechanism is faulty.
   2. Cassette housing strike against cabinet.

When cassette is placed in cassette housing, cassette does not lower

1. Cassette pop-up mechanism is defective.
2. Cassette housing strike against cabinet.

Automatic stop is not effected after auto rewind

1. Auto switch or repeat switch is defective.
2. Mechanism control is faulty.
3. Wiring is open between mechanism control connectors.

Main motor does not turn

1. Wiring to main motor is open.
2. Phase-advancing capacitor is defective.
3. Main motor is defective.

When play button is pressed, play motion is not effected

1. Loading end switch is defective.
2. Wiring is open.
3. Noise filter for DC motor is open.
4. DC motor is defective.
5. Wiring to DC motor is open.
6. Belt for DC motor is broken.
7. Belt has come off DC motor.
8. Belt from DC motor slips.
9. Belt for linking idler pulley and idler arm assembly is broken or has come off.

Capstan is not turning

1. Relay belt has come off.
2. Capstan belt has come off.
3. Belt slips.

When stop button is pressed during play or recording mode, tape is not taken up into cassette

1. Take-up reel torque is insufficient for unloading.

Stop takes place automatically at other than tape end while tape is run

1. Incident light is too bright (dim illumination or change light source position).
2. Tape has a scratch through which light passes (block the scratch with splicing tape).
3. Counter switch is defective or wiring is open.

Automatic stop does not take place at tape end during tape run
1. CDS is defective or wiring is open.
2. Cassette is defective (cassette has no transparent leader tape at tape end).
3. CDS lamp is dim.
4. Tension pole of mechanism is malpositioned.

"Still" cannot be applied
1. Pinch roller has not released the capstan shaft.
2. Weak back tension.

Picture drifts in still mode
1. Head drum rotation is abnormal.

Tape entangles in mechanism
1. Tape tension is wrong.
2. Tape at the pinch roller may be out of position.
3. Cassette compartment may be misaligned.

Tape moves too fast
1. Control track is missing.
2. Servo problem.

Tape damaged by machine
1. Improper tape threading.
2. Bad tape.
3. The ring is faulty and in need of adjustment.

Video heads do not rotate
1. Main motor dead.
2. Drive belts broken.

Tape transport is normal but VTR makes unusual noise

A. During play mode:
   1. Pole before pinch roller (wipe inside and bottom of rotary portion with oil dampened cloth).
   2. Between motor pulley and reel drive belt (wipe the belt with alcohol or replace it).
   3. Resonance noise of motor bearing (replace main motor).
   4. Resonance noise of relay pulley bearing (adjust pulley position so as to increase pre-pressure, or replace bearing).
   5. Slip between tape-up idler assembly and disk (replace rubber of idler).
   6. Resonance noise of disk spring (fix spring).
   7. Tape squeak (resonance of tape and tension arm: clean pole in cassette and guide pole of transport system or replace tape).
   8. Resonance noise of drum bearing (replace lower drum assembly).
   9. Noise of spring plate assembly–spring section resonating with drum rotation (replace bracket).

B. During loading or unloading mode:
   1. Roller and ring of idler arm (lubricate them).
   2. Pole before pinch roller (wipe inside and bottom of rotary portion with oil dampened cloth).
   3. Belt noise at time of unloading completion (adjust position of microswitch).

C. During fast forward or rewind:
   1. Counter pulley for relaying counter belt (lubricate them).
   2. Cassette and reel rubbing noise–caused by deviation of reel center (adjust position of positioning stud).
   3. Slip between motor pulley and belt (clean or replace belt).
   4. Rubbing noise between take-up idler and reel disk–abraded rubber tire of take-up idler (replace rubber tire).
   5. Clatter of take-up idler–take-up idler seized or need-

ing lubrication (clean and lubricate, or replace take-up idler assembly).

Machine will not play tape from another machine

1. Tracking control adjust.
2. Need servo alignment.
3. Control heads dirty.
4. Motor brake problem.
5. Tape tension.
6. Belts old or defective.

Picture not sharp

1. TV or monitor out of focus.
2. Camera out of focus.
3. Loss of high-frequency response.

6 • *Tape Maintenance*

Video Tape Structure

All video tape consists of three basic elements: a clear polyester base, a binder, and magnetic oxide particles.

The oxide particles play a major role in the video recording, since it is this part of the tape that actually contains the record of image and motion. Oxide particles differ according to type, size, and shape. The best oxide particles are hard and small, and come closest to a perfect needle shape. The smaller and more perfectly formed the particles, the more easily they can be equally dispersed in a given section of a video tape, thus providing better overall picture quality.

Evolution of Video Tape

In 1971 Sony introduced and marketed the widely successful three-quarter-inch U-Matic video cassette system. Sony took the next step in video evolution in 1975, introducing a 6⅛-inch long, 3¾-inch wide, and 1-inch thick, 60-minute high-density color video cassette smaller than an ordinary paperback book.

To fit more video information into a smaller video cassette, three factors were required: a shorter recorded wavelength, a narrow tape width, and thinner magnetic tape.

Conventional video recording systems place guard bands between video tracks, to prevent the signal leakage known as "crosstalk" from causing deterioration in picture quality. Unfortunately, guard bands take up space in video tape. So the azimuth recording system was created, which eliminated the need for guardbands while still maintaining a high video quality. The Betamax system uses two rotating heads: the azimuth of the cap on these heads is 7 degrees in opposite directions from the perpendicular, making a total difference of 14 degrees.

The video signals on each track are recorded at a slant, according to the azimuth of the head. During playback, the Betamax heads retrace their tracks. Should a head with different azimuth trace over the wrong track, no signal is reproduced. As a result, luminance crosstalk between adjacent tracks during playback is eliminated, and more video information can be recorded on a narrower tape width.

The success of this format led many other manufacturers, such as Sanyo, Sears, Toshiba, and Zenith, to produce a Beta-format video machine and tapes.

Care of Open Reel Tape

1. Handy small reels without large center holes should be held in a manner similar to that generally recommended for phonograph records.
2. Don't hold the reel in such a way as to squeeze the flanges against the edge of the tape which will possibly cause edge damage.
3. Use the hub of the large reels to hold the tape.
4. Don't smoke or eat around the tape. Particles can settle on the tape surface and cause dropouts.
5. A tape on fast forward or rewind has built up a momentum which, if suddenly stopped, will cause it to buckle, grinding any loose debris on the tape into the oxide surface. This is called "cinching."
6. Don't leave the tape on cued position for a long period of time, allowing air particles to settle on exposed tape.
7. Don't splice tape or put any gummer tape on the video tape.

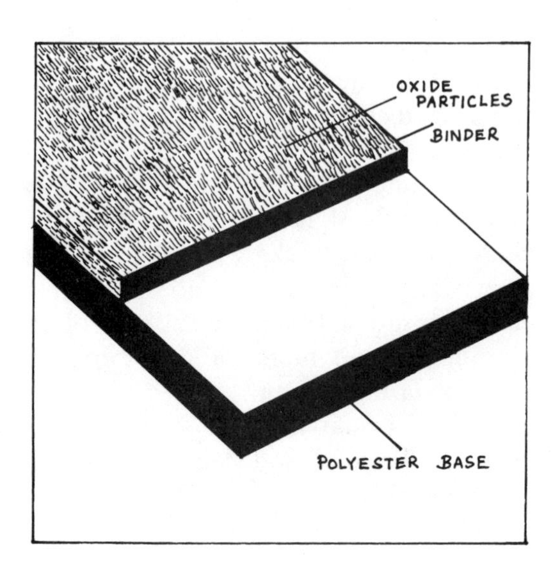

A cross section of a video tape

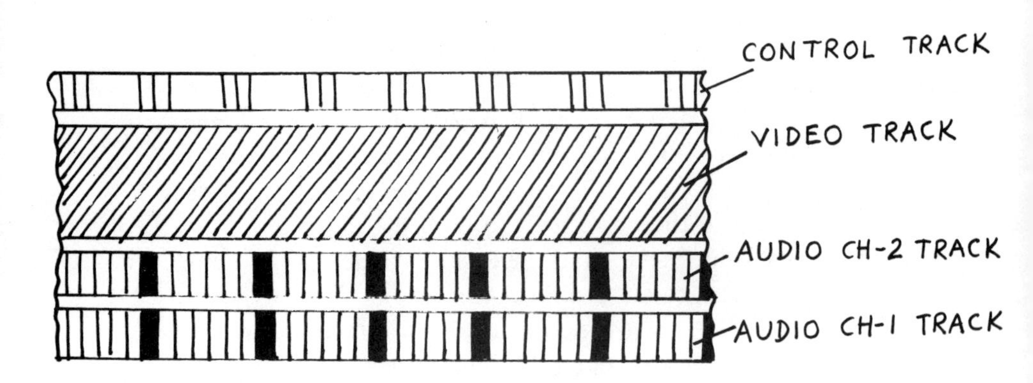

A cross section of a ¾″ U-Matic video tape

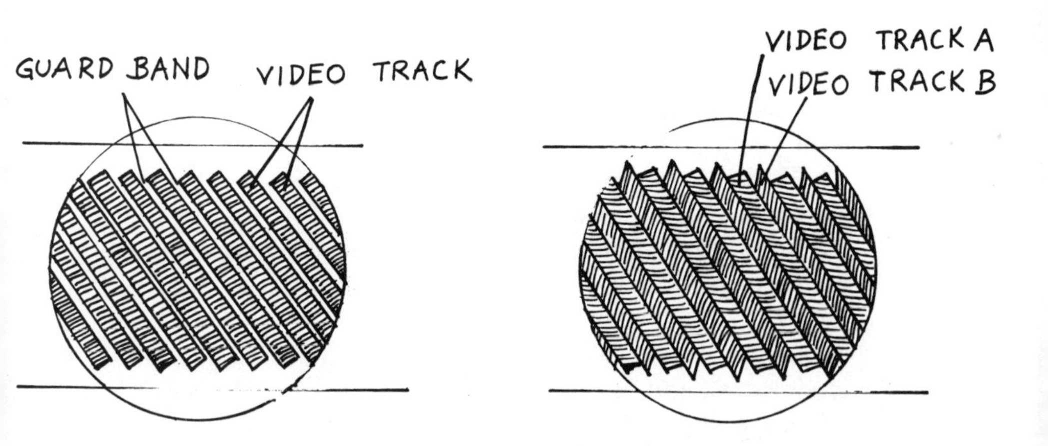

A comparison of the Ozimuth (right) and conventional video recording systems

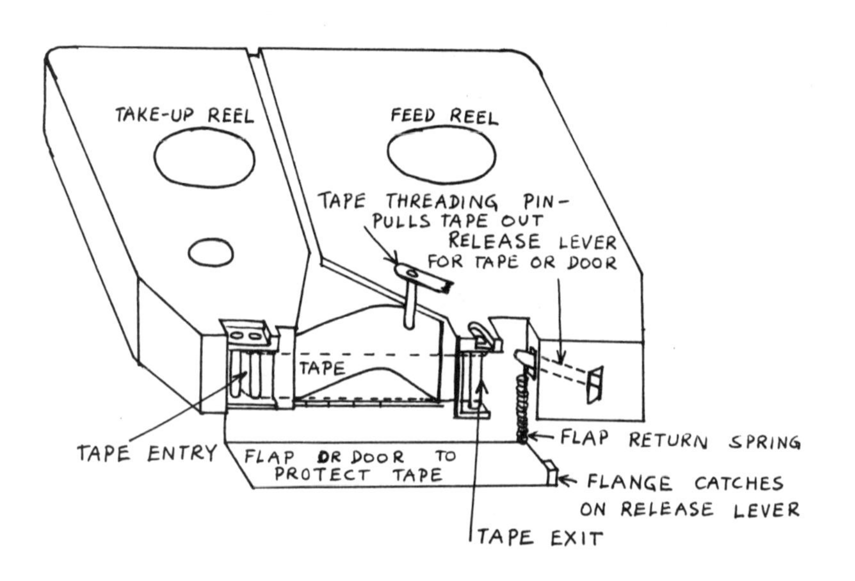

Cassette with flap or door open

Correct Use and Handling of Video Cassette Tape and Reel

1. Reel and cassettes should be stored upright in their protective cases, rather than on a flange. If a reel is left lying horizontally the tape pack can shift, causing the tape edge to rub on the bottom flange, probably damaging the tape.
2. Resist the urge to tinker with a malfunctioning cassette. By opening and closing a cassette, you are likely knocking it out of tolerance. By removing and replacing a screw you are also creating a fine plastic dust which can get onto these tapes and cause dropout.
3. Keep the video cassette away from high temperatures, excessive dust, and moisture.
4. Keep the video cassette away from strong magnetic fields.
5. Do not leave the video cassette exposed directly to the sun.
6. Before inserting, take up slack in the tape by turning the reels manually in the direction indicated by the arrows.
7. Avoid repeated insertion and removal of the cassette without operating the machine.
8. If the video cassette has been exposed to cold temperatures (68° F or below), it should be stored at room temperature for at least one hour before use. Condensation forming on the tape will cause the tape to stick to the video drum and result in damage to the tape.
9. Do not store your video cassettes in the trunk or glove compartment of your car.
10. Do not touch the tape with your fingers, since the oils from your hand will affect the operation of the machine. Also the oil on the tape will touch the video head and cause dropout.
11. Avoid violent vibrations or shock. Do not drop the cassette.
12. The video cassette is equipped with a safety tab to prevent accidental erasure. Remove the tab when you wish a recording to be protected from accidental erasure. If you wish to record on a cassette having the tab already removed, place a piece of cellophane tape over the hole.

13. The video cassette tape should not be spliced. A splice on the tape could cause damage to the video heads.

14. Use sensitive stickers for easy identification of your cassette.

15. Always rewind your tape after showing.

16. Do not leave your tape in the machine overnight.

17. Always store and ship the cassette inside its plastic protective case.

18. Do not put it on top of airconditioners or heaters.

19. Do not pull the tape out of the cassette. This invites damage and shortens its life.

20. If the tape has rewound unevenly, rewind it again to pack the tape properly.

21. Do not place a recorded cassette next to an electric motor or power transformer; the magnetic field around them might accidentally erase the tape.

Copying: Tape to Tape Duplicating

Copying one tape onto another tape is one of the easiest things to do with your VTR. You must employ two video machines to perform this function. One video machine will perform as a playback unit; the other machine will operate as the recorder.

A video cable is inserted into the video input of the playback machine, and the other end of the video cable is inserted in the video input of the recorder. The same procedure goes for the audio cables.

The audio output of the playback machine goes into the audio input of the recorder.

Remember that the playback unit will house the master tape, and the recorder will duplicate the master tape.

If your video cassette recorder does not have an Automatic Gain Control for your audio, then you must employ a Sony ALT-1 audio line transformer between the two machines. This will extract any hissing in the audio track.

Another way of duplicating a videocassette is to use your RF out signal, and a 75-ohm coaxial into the RF out signal of your playback machine, and plugging the other end of the coaxial cable into the VHF input signal on the record machine. This will eliminate the video cable and audio cable. The selector switch on the VTR must be set to "TV" to perform this operation.

7 • *Video Beam Projector*

Projected TV: The Big Picture

Big-screen projection television has been around since 1949. For many years companies like General Electric, Projection Systems, and Eidophor have been making equipment available for presentations to large audiences. But the price of these systems—$25,000 to $100,000 just for rental—pushed them out of the reach of most potential users. Projection television thus became the domain of fight fans and the rock scene, carrying live coverage of boxing matches or concerts to audiences around the globe.

In 1973 Henry Kloss, an audio pioneer who spent his adult life dreaming up ways to turn technology into playthings for grownups, began marketing the Videobeam, a three-color-tube projection system with a 6-by-4-foot screen, that produced a brighter picture, more than ten times larger than that of ordinary TV sets, for the extremely low rental price of $2500. Advent was not the first with a two-piece projection TV unit. Sony beat them to the market with a similar two-piece unit selling for more than $3000.

Regardless of what came before him, Henry Kloss sold a hundred of his units from a tiny showroom in the back of his Massachusetts warehouse without advertising and before officially introducing it to the national market.

Many an entrepreneur recognized the general potential of big-screen TV, and started to manufacture and market these low-price giant screens. One of the results was Muntz Home Theatre Vision, with a price of $1595. Its feature was a 40-by-

32-inch Kodak Ektalite screen. The unit uses a modified Sony Trinitron to project a picture through a single lens to a mirror and into the screen.

Many imitators, using a small, current-model TV and a single lens and mounting an Ektalite screen in various kinds of cabinets, have been hitting the market, selling their versions below $800. And still other imitators are using a cardboard cutout attached to a mirror and lens that throws a picture on the wall, at the still lower price of $49.

Large-Screen Video Projection

The Advent VideoBeam System

Color television broadcasts are a mixture of three color signals—green, blue, and red. When the signals are properly recombined in an ordinary color television set, the full range of visible color is produced.

However, there are *multitube* video projectors, which employ three separate light projection tubes, one for each of the three color signals. Inside each of the trio of tubes is an electron gun for the color the tube will transmit to the screen across the room. The gun is aimed to sweep back and forth over a 3-inch phosphor-coated target screen inside the tube, which lights up in the color of the phosphor coating.

Since this light-emitting target need produce only one color, it can be uniformly coated with a single phosphor for the desired color: The color dots or lattices of conventional sets are not needed, nor is a show mark, thus providing far higher efficiency than possible in a direct-view picture tube. The accompanying illustration shows how the picture from the light-emitting target is projected. The light given off is reflected by a spherical mirror, which directs the light out of the tube by means of a corrector lens.

The mirror focuses the light from each point in the internal target on the corresponding area of the big screen across the room. And when the separate red, blue, and green images from the three light guide tubes converge on the big screen, the full-color picture is produced.

As roundabout as this may seem in the description, it is a good deal simpler to execute precisely than the usual color television configuration. The beam from the single electron gun inside each guide tube sweeps back and forth over a span of only 15 degrees instead of the 90 degrees required by the design of ordinary picture tubes, making it far easier to maintain excellent picture linearity (uniformity) across the screen. With no need for clusters of color dots or for the color lattices employed in more recent picture tubes, and no concern that an electron beam may strike the wrong color phosphor, the usually fussy convergence of three colors on the face of a picture tube becomes a simpler matter of overlapping the beams from the three light-guide tubes on the big external screen.

The design of the light-guide tube is a critical factor in the performance of the system. The Schmidt optical system it em-

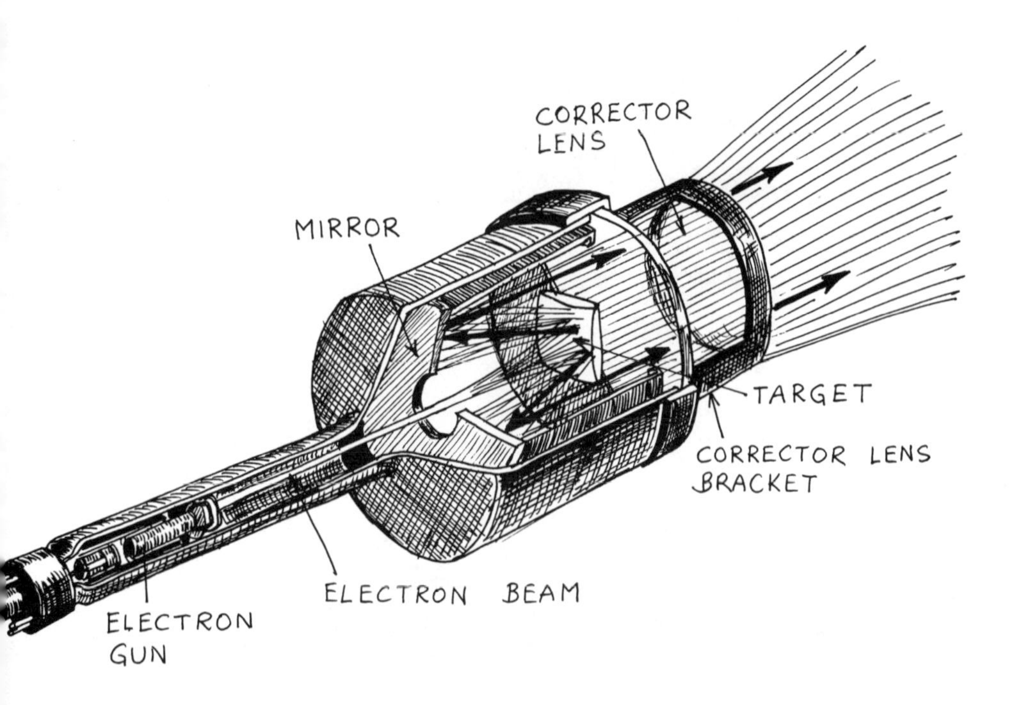

The Light-Guide Projection tube

The optical system of the Advent model

ploys was first used for wide-angle telescopes in observatories. In previous projection television systems for theaters and industrial use, the Schmidt or other optics were located outside the projection tube itself. They felt this was unacceptable for a system to be used in homes and elsewhere by non-technical people, since the external optics required careful operation and maintenance by a knowledgeable technician.

Advent designed (and manufactured) the Light-guide tube with all critical optical parts sealed in a fixed relationship inside the tube. This design allows all critical optical adjustments to be made once and for all in the manufacturing process rather than by the operator of the set, and it keeps the highly vulnerable optics out of the way of dust and damage. Only the front of the corrector lens is exposed, requiring occasional dusting. Also critical to the set's performance, and also designed and manufactured by Advent, is the large screen. It employs a new, highly reflective aluminum surface material (manufactured under license from Kodak) that is five times more reflective than a conventional lenticular or beaded screen. It is the combined efficiency of the new screen and the light-guide projection tubes (which have no shadow mask to absorb and waste potential picture brightness) that achieves the brightness necessary, for a satisfying picture in a room with background lighting. The efficiency of the VideoBeam projection system is so high that, despite the tremendous brightness requirements of a projection system, the set consumes less power than conventional vacuum-tube color TV sets and no more power than the latest solid-state color sets.

Maintaining Picture Quality in the Advent System

For any television set to produce its best picture quality over a long period of use, the set must have a means to adjust color convergence (the precision with which the three color signals merge to produce a full-color picture) and focus. In conventional television sets, however, these adjustments are too complex to be left to the user, and the result is that the set owner, who is understandably reluctant to call in a TV serviceman as long as things are working at all, tends to accept slow deterioration of the picture as inevitable with age.

Since Advent expected anyone owning a VideoBeam set to demand high picture quality, they felt it crucial from the start to supply a simple way for a user to keep color convergence and internal focus up to original standards—and to regain best-picture quality if the set should be moved or jarred. In addition to simple controls to achieve this purpose, the VideoBeam set provides a special switch that displays a rectangular crosshatch line pattern on the screen. The combination of simple controls and an easy-to-follow visual guide makes the adjustment process simple enough for the owner to check picture quality whenever he is in doubt about it for any reason.

The Advent VideoBeam System: Essential Facts

The role of the Advent System is important enough to warrant this summary of the facts:

Picture Size

6′ diagonal; 45″ h. × 60″ w.; 2700 sq. in. Over 8 times larger than a 25″ set.

Screen Characteristics

High reflectivity, uniform brightness from center to edge (no "hot-spotting"), broad horizontal viewing angles, and the ability to reject off-axis ambient light.

Optical System

Three tubes, each with a four-element lens system with f1.3 optical speed.

Brightness

Approximately twice the standard for movie theaters.

Anode Voltage
28.5 kV; moderate beam currents provide extended tube life and consistently small spot size without phosphor "blooming."

Electronics
All solid state; computer-controlled electronic tuning with keyboard-touch and LED channel display.

Remote Control
Five-function wireless remote control with random and sequential channel access; continuous volume; volume mute; and power.

Audio System
Wide-range speaker with active frequency contouring for high fidelity sound. Separate output for use with home stereo system, controlled by main or remote volume controls.

Interface
Direct video and audio inputs and recording outputs for connecting video recorders, video games, or home computers; provides better performance and eliminates need for RF modulation.

Antenna Inputs
300- and 75-ohm inputs for both VHF and UHF.

Accessory AC Outlet
Unswitched; 200 watts.

Cabinet Finish
Walnut veneer and hardwood solids.

Dimensions
Console: 17⅜″ h. × 32 13/16″ w. × 23½″ d. Screen: 46″ h. × 60″ w.; 66″ h. with legs; 72″ diagonal. Projection distance: 87½″. Total distance from back of screen to back of projector: 9′4″ (112″).

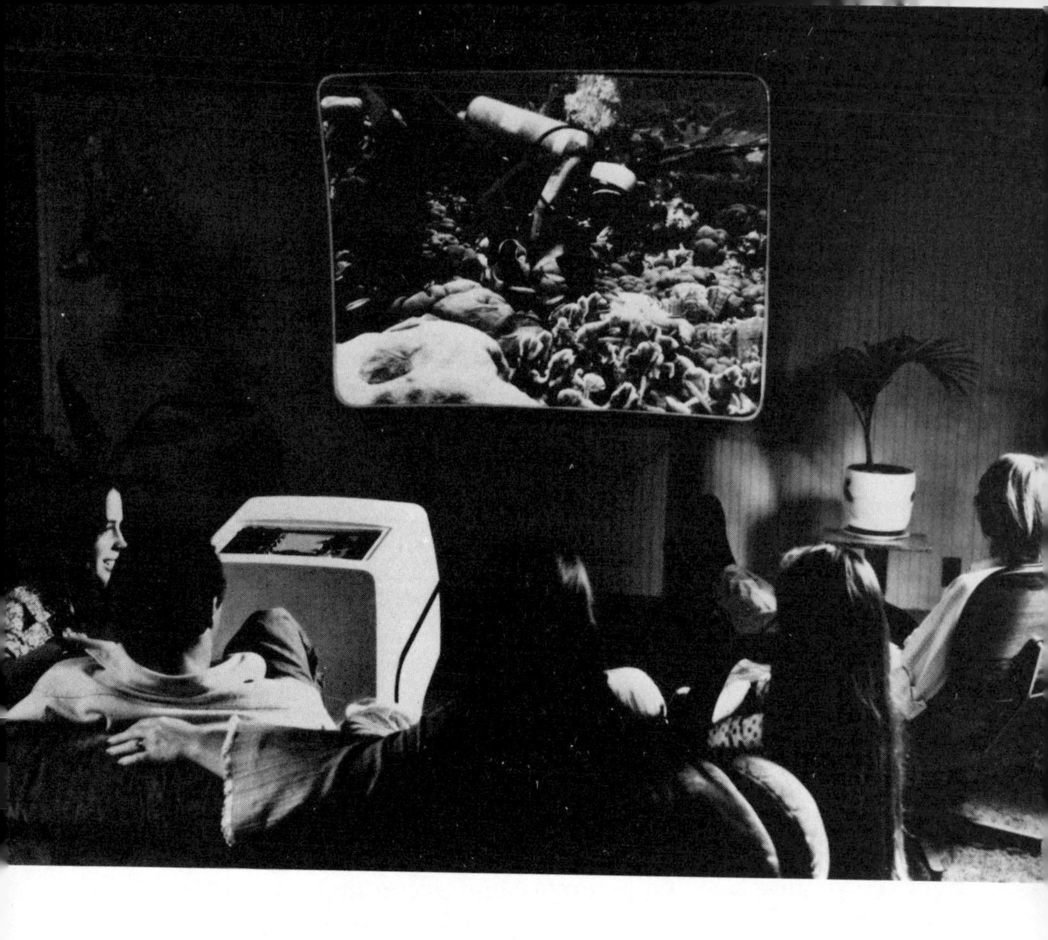

Four examples of the Advent VideoBeam projection

99

Questions and Answers on the Advent Video Beam Projector

Q. *Do I use my present TV with a VideoBeam set?*
A. No, it is a complete color TV itself. It comes complete with receiver/projector and screen. All you need supply is an antenna or cable connection—just like your current TV set.
Q. *Does it pick up all the channels?*
A. Yes, all the channels you get in your area, both UHF and VHF.
Q. *Do I need a special antenna?*
A. No, not a special antenna, but a good one. The big picture reveals ghosts and other signal flaws more than a regular set, so a good roof antenna, or a cable connection, is what you should have to fully enjoy the set's potential.
Q. *Does the set make video tape recordings?*
A. No, a VideoBeam set is a TV set like any other, so it can't make video tapes on its own. You need a separate video cassette deck for that.
Q. *Can I use my own screen, or project the picture right on the wall?*
A. No, it's necessary to use the screen that comes with the set. Performance, particularly picture brightness, would suffer without the special screen supplied.
Q. *How big a room do I need?*
A. That's entirely up to you. VideoBeam TV will fit into a room as small as about 10 by 12 feet, up to as big as you prefer.
Q. *Does VideoBeam TV use a lot of electricity?*
A. No, not at all. It uses about the same as a modern solid-state 25-inch color console (a maximum of 175 watts of regular 120-volt household current). That's half or less the power consumed by the all-tube color sets of just a few years ago.
Q. *Do I have to turn off all the lights to watch?*
A. No, just don't have a source of bright light centered right on the screen. Light from the sides, or above, will affect the picture very little. And when you're seated in front of the

screen, the picture is considerably brighter than a movie in a theater, and much brighter than home movies.

Q. *Is VideoBeam TV harder to operate than a regular set?*

A. No, its regular operating controls work just like those on a conventional color set. And the special controls that let you touch up color sharpness (unlike any other set) will take you but a few minutes to learn to use, and thereafter just a few seconds to operate when needed.

Q. *Is there any greater danger of X-ray radiation than with a regular set?*

A. No, don't be misled by the fact that there's something going between the projector and screen; that's just colored light, like a movie projector. Furthermore, the big, bright picture is created without using higher voltages inside the set than those of a regular 25-inch set. And most importantly, each VideoBeam set meets the same stringent federal standards for radiation that all color sets must now meet.

CV3 Projector for the Consumer

Projection Systems, Inc. (Clifton, N.J.), a company that has been in the business a long time, manufacturing projector TVs for large audiences (viewing mostly rock concerts and boxing matches), has come up with a projector system for the connoisseur consumer, priced at $6000.

The CV3 projector can receive live television broadcasts, closed-circuit transmissions, or video cassettes in the normal way. Where it is different from other projectors is in its use and refinement of the Schmidt optical system, and in the solid-state circuitry that gives a high light output and at the same time makes possible projection onto a flat screen.

Moreover, the design arranges all the working parts into a very compact unit, measuring 33 by 29 by 12 feet and weighing 100 pounds. The size and weight allow users to place the projector where it is most convenient—either suspended from the

Floor model CV3 video projector

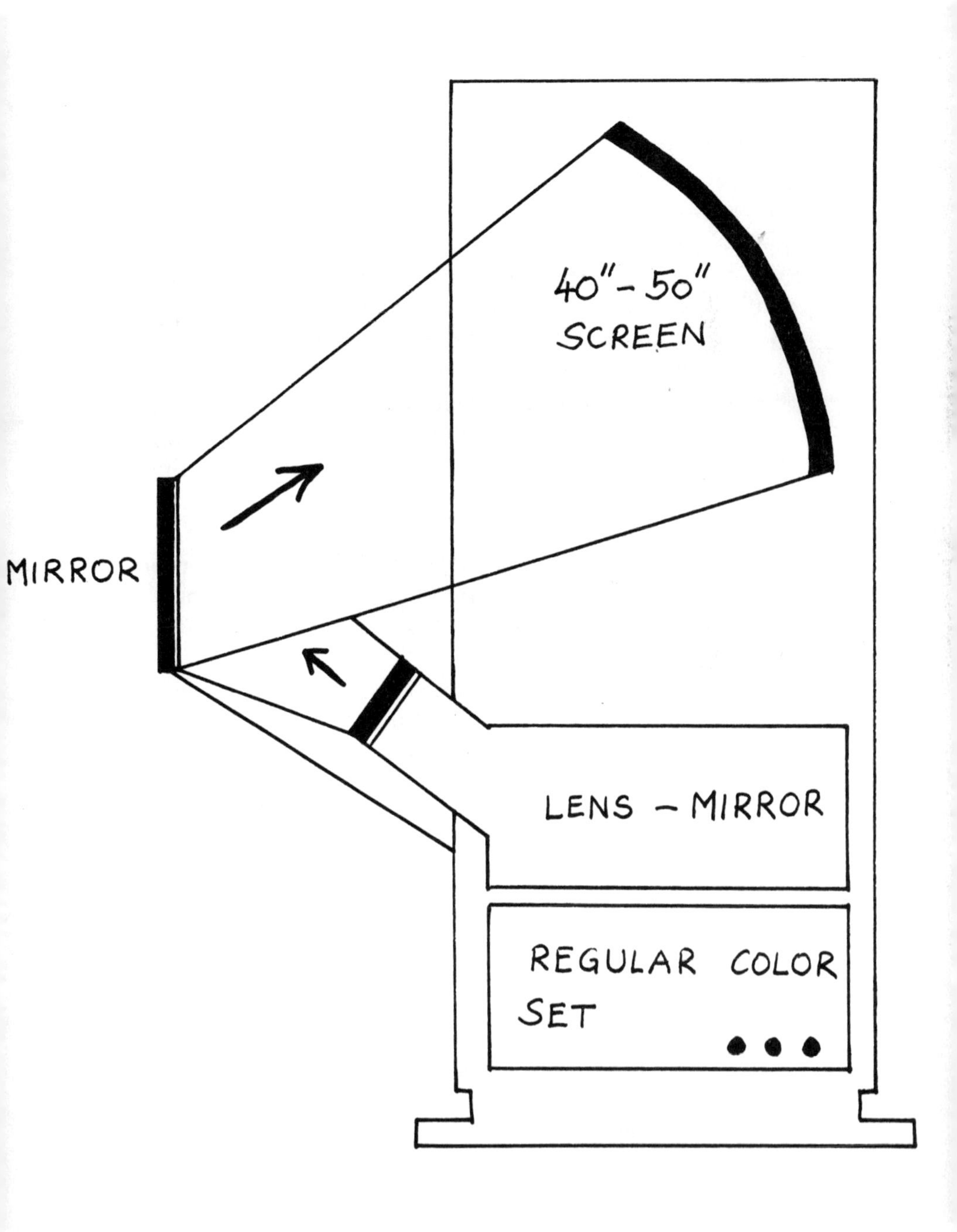

Diagram of a single-tube projector

ceiling or standing on the floor in the conventional way. A standard remote-control box makes the operation of the projector from either type of mounting a simple affair.

Like most other color television projectors, the CV3 has three projection tubes, one for each of the three primary TV colors. Reflecting through a combination of mirrors and lenses, these tubes project three separate beams of light that converge on the screen to form a perfect color picture. The picture is bright enough for excellent viewing with limited background lighting. And because the three primary colors are beamed independently, the picture does not have the color dots or stripes seen on many conventional color television sets; in fact, it bears a close resemblance to the picture projected onto a cinema screen. The live-broadcast picture, however, can be only as good as the signal received. Reception flaws are magnified on the CV3 superscreen, so if perfect viewing is to be maintained at all times, a strong signal and a good antenna or cable connection are essential. There is no radiation hazard: only light beams travel from the projector to the screen. The projector unit includes a 3-watt internal speaker system; when extra sound volume is required additional external speakers can be linked in.

The control box can be attached to the projector or operated remotely. It has been designed so that the ordinary individual can adjust the CV3 in much the same way as a standard color television set. The controls include:

Power on/off
Sound volume
Color
Brightness
Contrast
Enhancement (sharpness of image)
Crosshatch generation (alignment, normal/test)
TV/VTR selector

The single-tube units all utilize a refractive-type system in which the output of a standard picture tube is magnified through an external optical system.

Many of the early single-tube projection systems used the

high-gain Kodak Ektalite screen, which was very sensitive to rough handling. The newer screens that have been developed resist fingerprints and are washable. If you want to keep your screen dust free, it will have to be cleaned periodically.

Ankar's RM64 Projection TV

In addition to Advent's three-gun system, a new one-gun projection TV has been introduced by Ankar Video, a New York-based corporation located in Merrick, N.Y. In the past, no one-gun system was ever considered since all models suffered the same problem: A dull, washed out picture. This was because of the insufficient light emanating from one picture tube. Ankar has solved this problem by use of a F-1.3 lens and a highly reflective aluminum based screen. The picture I viewed was judged to be almost as bright as the Advent's. This system consists of four parts: A 12″ Quasar Color TV, a walnut-veneered cabinet, a two element F-1.3 lens and a plastic-coated aluminum-based screen that can be either hung or free standing. The console measures 23″ × 19″ × 33″ (h/w/d) and weighs 55 lbs; the screen measures 40½″ × 50½″ (h/w) and weighs 20 lbs. The suggested retail price is $1050.

The system arrived in three boxes and took approximately 30 minutes to assemble. The TV is placed into the cabinet and hooked up to the antenna. The screen is then hung at the desired viewing height. Cabinet screws are adjusted to raise the cabinet to the height of the screen. The lens is placed in the cabinet and adjusted to a sharp focus.

The picture is produced by magnifying and projecting the TV image through the lens onto the screen. Since this system was so similar to Advent's, a comparison of the more important factors appears in order.

On the negative side, I found that Advent's three-gun system gave us a slightly brighter picture than Ankar's and therefore, the screening room do not have to be dimly lit. Advent's audio quality was judged somewhat richer and deeper. (Advent uses a 5″ high density dynamic speaker as opposed to the 3″ speaker

Front and side views of the Ankar Projection TV

of Quasar's 12″ TV). In addition the Advent was judged to be a slightly better looking unit.

The positive attributes seem to outweigh the negative ones. Ankar's screen is washable and plastic coated. This makes it extremely resistant to any damages. Advent's screen is extremely delicate (sunlight, dust and neglected handling will cause damage). Also, since Ankar's screen is much lighter it can be hung like a mirror and does not take up the space of a free-standing screen. It was also judged much easier to focus Ankar's picture since only one picture tube is used. The lens slides in and out until the perfect focus is achieved.

Advent's three lightbeams arrive at the screen from different angles. If the beams are not properly aligned, it's extremely difficult to realign them. Also, moving slightly to the right or left of Advent's screen, causes a shift in color or tint. Ankar's system has a 140° viewing angle with no change in tint or color. The simplicity of Ankar's unit is another important factor. Since it uses only one gun, it's less costly to produce and therefore retails at $1050, instead of $2595. The 12″ Quasar TV inside can be removed and serviced the way you would a normal portable. There are no costly service charges. Also, since only one tube is used, it costs no more to operate than a 12″ portable.

In conclusion, while both systems have their faults and merits, the choice of a system rests with an individual's judgement of audio and visual priorities.

The Ankar's screen is washable and plastic coated.

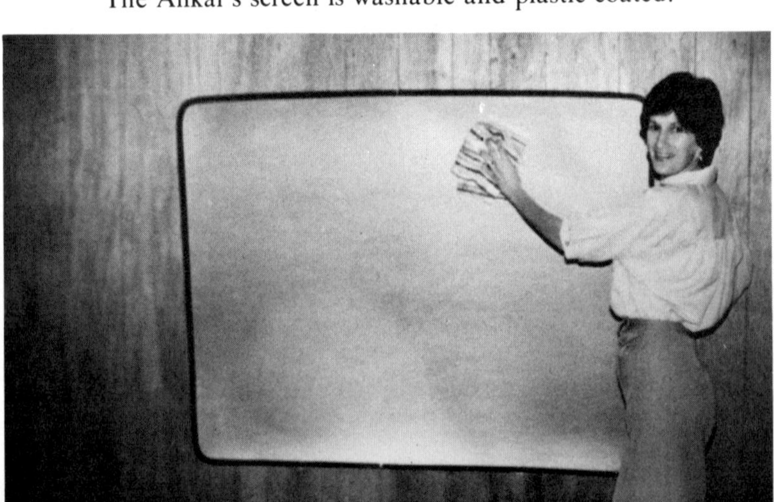

Projection System Characteristics

Make/Model	*Screen Length/ Width (in.)*	*Screen Size (Diag.)*	*Screen to Proj. Distance*	*Remote?*	*Remote Function*	*Video Input*	*Audio In/Out*
SINGLE-TUBE PROJECTORS							
Projection TV of America	32 × 40	52″	72″	wired	channel	yes	yes/yes
Sega/Sega-vision CR-511	32 × 40	50″	NA	wireless	channel	yes	yes/yes
Sony KP-4000A		40″	NA	wireless	channel on/off/vol	yes	yes/yes
Spearsonic Elec/Viditron		64″	52″	wired	channel	no	yes/yes
Spectra-Vu	32 × 40	51″	66″	wired	channel	yes	yes/yes
Steck Mfg. Sigma 51	32 × 40	51″	48″	opt. wireless	channel on/off/vol	yes	no/yes
Superscreen Television	32 × 40	50″	NA	wireless	channel on/off/vol	yes	yes/yes
Tandom/VPS-3	32 × 40	50″	varies	wireless	channel on/off/vol	yes	yes/yes
Theater Vision	32 × 40	50″ up	varies	wireless	channel	yes	yes/yes

Built-In Speakers	*External Speaker Connection*	*Test Signals*	*Washable Screen*	*Type of TV Chassis*	*Warranty Parts/ Labor*	*Projector Weight*	*Price (Dollars)*
yes	yes	no	yes	12″ Quasar	1 yr/90D	80	1300 up
yes	yes	no	yes	Sony	1 yr/90D	300	2195 up
yes	NA	no	yes	Sony	1 yr/90D	225	2295 up
no	yes	no	yes	Quasar	1 yr/1 yr	40	NA
yes	yes	no	yes	Sony	1 yr/90D	75	2860
no	yes	no	yes	13″ Toshiba	1 yr/90D	25	995
yes	yes	no	yes	Sony	1 yr/1 yr	NA	1495
yes	yes	no	yes	Quasar	1 yr/90D	46	1595 up
yes	yes	no	yes	NA	1 yr/1 yr	125	995 up

Projection System Characteristics

Make/Model	*Screen Length/ Width (in.)*	*Screen Size (Diag.)*	*Screen to Proj. Distance*	*Remote?*	*Remote Function*	*Video Input*	*Audio In/Out*
Video Ind. Videomaster 450	32 × 40	52″	NA	wireless	channel on/off/vol	yes	no/yes
Videorama Ultra 2000		48″	6′	wireless	channel	yes	yes/yes
Viewpoint V-2000	32 × 40	52″	NA				NA
Voorhies Ent. Starbrite	32 × 40	52″	NA	wireless	channel on/off/vol	no	no/no
MULTITUBE UNITS							
Advent/1000A		84″	100″	opt. wired	chan/col on/off/vol	yes	yes/yes
Advent/750		72″	87″	wireless	channel on/off/vol	yes	yes/yes
Advent/710		60″	72″	no	none	no	no/no
Mitsubishi MGA/VIDEOscan		72″	72″	wireless	channel on/off/vol	yes	yes/yes
Panasonic CT-6000		60″	NA	wireless	channel on/off/vol	yes	yes/yes
Projection Systems/CV3	72 × 96	10′	varies	wired	channel convergence	yes	yes/yes

A summary of many aspects of 19 projection models. Note that many of these—unlike those systems discussed in the text—involve single-tube types. NA—Not available

Built-In Speakers	*External Speaker Connection*	*Test Signals*	*Washable Screen*	*Type of TV Chassis*	*Warranty Parts/ Labor*	*Projector Weight*	*Price (Dollars)*
yes	yes	no	yes	Sony	1 yr/1 yr	175	1595
yes	yes	no	yes	various	90D/30D	36-76	550 up
NA	NA	no	yes	various	1 yr	64	599
no	opt.	no	yes	Zenith	1 yr/90D		
yes	yes	CH	no	Advent	1 yr/1 yr	140	3995
yes	yes	CH	no	Advent	1 yr/90D	95	2995
yes	no	CH	yes	Advent	1 yr/90D	95	2595
yes	yes	CH	yes	MGA	1 yr/1 yr	193	3400
yes	yes	CH	yes	Panasonic		180	
yes	yes	CH	yes	PSI	1 yr/90D	125	6000

8 • *Video Disc*

The Historical Background of the Videodisc

When the video cassette boom got under way in the early 1970s, several companies tried a different approach in the field of video and developed the Videodisc. In the years that followed, a few companies dropped the idea, but not RCA or Phillips/MCA. In the early stages, MCA had started its own development of a system called "Discovision," which was somewhat similar to a Phillips version. In the interest of both companies, and the consumer, a common encoding and Videodisc format was adopted to achieve interchange ability of consumer optical Videodiscs. The best technology of both companies was used to achieve optimum performance inherent in both the Videodiscs and players. The adopted formats were reviewed with other major electronics corporations and, as a result, the encoding and disc formats and the specifications were made available to United States industry.

Phillips/MCA Videodisc: Structure and Manufacture

The Phillips and MCA system is a noncontract optical system in which a prerecorded Videodisc is used in conjunction with the Videodisc player. A standard TV receiver may be used for the audio and video. The TV can be either color or black-and-white, of any size or make.

One player can be connected to a number of TV receivers. If many sets are used, it may be desirable to use a line amplifier as well.

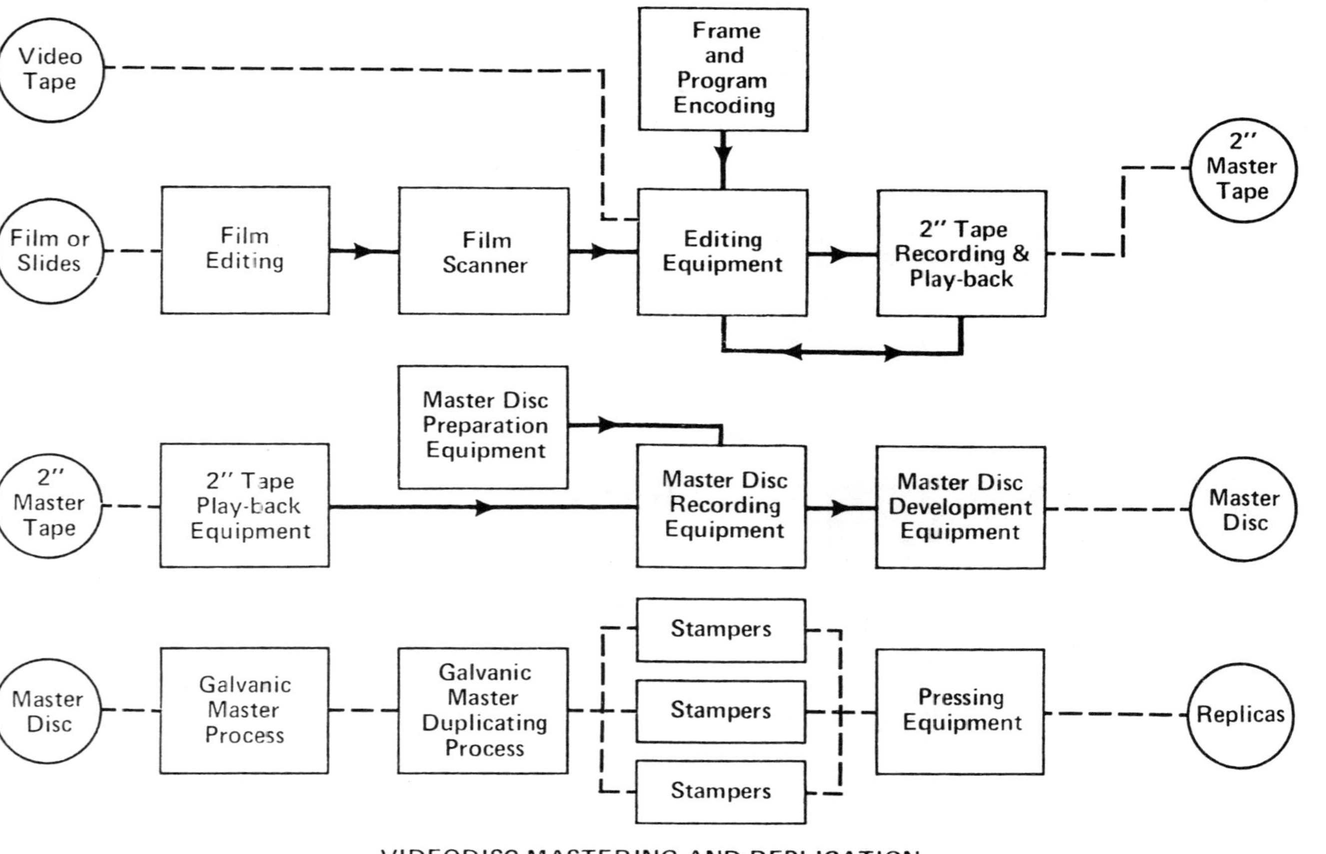

VIDEODISC MASTERING AND REPLICATION

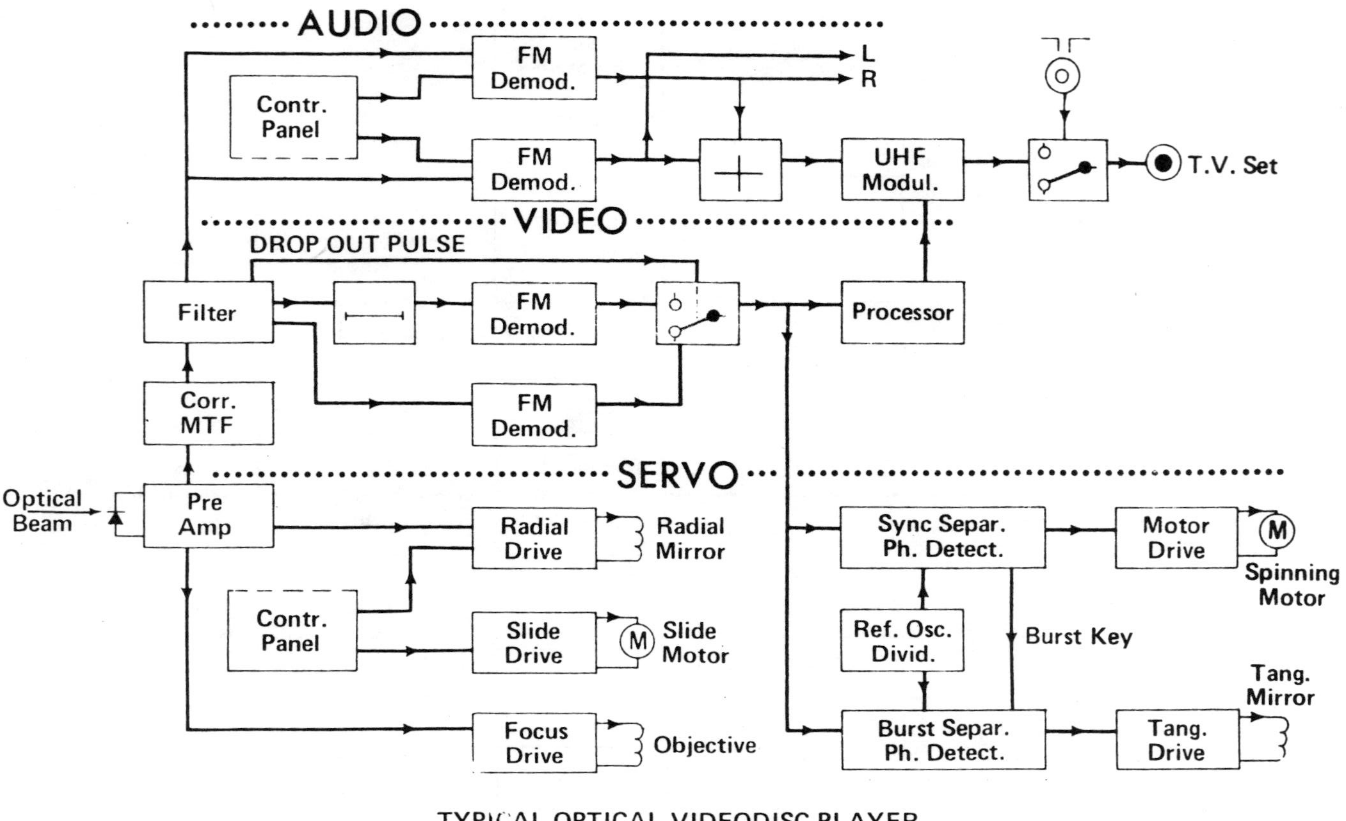

TYPICAL OPTICAL VIDEODISC PLAYER

Video tape or film is the input for the technical editing process. One- or two-inch video tape and 35- or 70-mm film are the preferred inputs. However, three-quarter-inch tape and 16-mm film are acceptable. The technical editing equipment encodes lead-in track, the frame numbers, the chapter stops, the automatic frame stops, and the lead-out track. This information is then transferred to a 2-inch tape that is used as the input to the master recorder.

The master disc is recorded in real time. Metal stampers are made from the glass disc master for replication of the discs. There are two methods of high-volume replication: injection molding and compression molding. A low-volume replication method is available with the photopolymerization (or 2 P) process. Videodiscs of various diameters (as on the following list) will also be available that can all be used on the same player:

(1) 12″—30 minutes per side disc.
(2) 12″—60 minutes per side disc.
(3) 8″—13½ minutes per side disc.

The difference between an 1800-RPM 30-minute disc and a 60-minute disc is in the spacing and length of the pits and the variable rotational speed of the 60-minute disc.

The pit configuration for a 60-minute disc is identical in the inner and outer diameter. Each track contains varying numbers of TV frames. The disc speed starts at 1800 RPM and slows to almost 600 RPM at the outer diameter. With the one-hour disc, the search forward and reverse will still be used for entertainment programs. The 30-minute disc contains 54,000 TV frames, each frame being one revolution of the disc. Each picture is defined by the vertical blanking interval. The playing time is determined by how the individual wishes to view the program, giving complete viewer interaction.

A reflective aluminum coating is evaporated over the upper surface of the disc, and over this is placed a protective plastic coating. The light beam, reading the information from underneath, passes through the disc, picking up the information, which is reflected back down through the optics. Any fingerprints, dust, or scratches are out of the focal plane of the light beam. Because the information is read from the underside,

The Videodisc System. Its features are described in the text.

the upper surface is not critical. This feature eliminates the necessity for meticulous handling of the discs.

The Videodisc player is connected to the TV antenna leads in the same way as the box attachment for video games. Either channel 3 or 4 may be selected on the TV set, whichever is not in broadcast use in the area.

The optics module contains tracking mirrors and a special prism. The laser and optics module move radially on a mechanical slide. This allows for all the desirable features, such as freeze frame, slow motion, and the like.

When using freeze-frame, the same track is being reread by the light beam. At the vertical blanking interval when the beam would normally begin to read the next track, it jumps back and rereads the same track and will continue to do so until you move it by selecting another feature.

Videodisc Features

The accompanying photograph shows the Videodisc player and its features. Reading from left to right, these are:

Power–on and off
Still frame–forward and reverse (moved manually)
Slow motion–forward and reverse
Slow motion rate slide–for speed variation: allows for one frame every four seconds, increasing almost to real time. This feature can be used as a browsing mode for still frames or viewing an action scene in slow motion.
Main control–Play, forward, and reverse (real time)
Fast forward–3 times normal speed or an apparent 90 frames per second.
Audio I and II–Two discrete audio channels allowing for stereo or any two narratives with the same video, such as in two languages, in adult-child versions, or in technical and non-technical lingos.
Visual search–Rapid forward and reverse. The light beam scans the disc, either forward or reverse, in 15–20 seconds. It

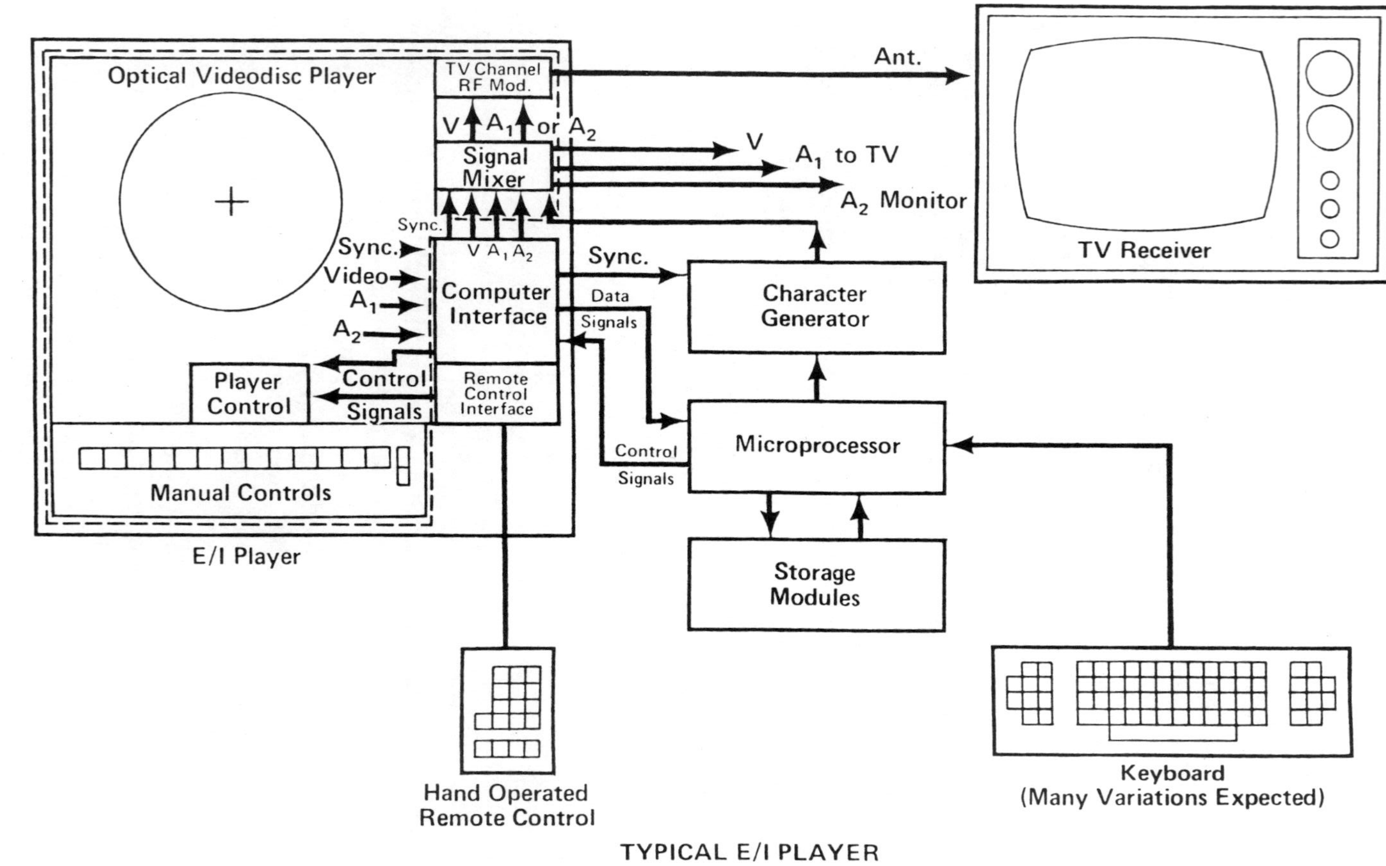

TYPICAL E/I PLAYER

randomly picks up images about every 400 frames. Using the index number as a guide, any section of the disc can be located quickly. If the disc is coded for chapter stops, the search button when chosen in either forward or reverse will cause the light beam to automatically stop and lock in on the next frame with a chapter stop code. Seventy-nine chapter stops can be encoded on the disc.

Index or frame no.—Button to the far right. The frame number is encoded on the disc itself in the vertical blanking interval. By pushing the button, it can be called up or taken out, as you wish.

The combination of these features plus the strategic placement of the chapter stops, automatic frame stops, and frame visual instruction allows for a fully interactive system for manual branch programming.

Discs are of about the same size and thickness as today's long-playing audio records but are generally more rugged when it comes to handling, storage, shipping, and use. Some discs will be flexible and can be bound into print publications or rolled up and mailed in a tube.

Major catalog sales chains are now planning projects which would use a Videodisc itself as a catalog to show the customer the item and then automatically show it in operation, along with a sales pitch audio track. The throw-away audio record you see bound in with the pages of publications today will be a Videodisc tomorrow.

The Videodisc albums will retail at a price slightly higher than an LP record, from $2 to $25. Videodisc albums will cover a wide range of entertainment: sports shows, cooking lessons, encyclopedic discussions, and, of course, full-length feature films. The MCA lineup includes blockbusters such as *Jaws, The Sting, Earthquake, MacArthur, Airport,* and *The Hindenburg,* as well as classics such as *Animal Crackers, Going My Way,* and *Frankenstein*—three hundred titles in all.

Specifications for the Phillips/MCA Optical Videodisc System

Unique Characteristics of the Videodisc System

1. Inexpensive carrier. The disc is made of low-cost material and lends itself ideally to high speed and mass replication.
2. Contactless reading avoids wear of the information carrier and reading device, guaranteeing continued good picture quality.
3. Protection of information facilitates handling of the disc, since quality is unaffected by dust, fingerprints, etc.
4. Playing time up to 30 minutes for a 12-inch diameter disc.
5. Random access to any part of the program.
6. Still pictures–max. 54,000 on a 12-inch diameter disc.
7. Fast forward and slow motion.
8. Coding possibilities for automatic action such as access, start, stop.
9. Individual identification of each frame by means of a visual display.
10. Two discrete channels for stereophonic reproduction or two languages (or other uses, as previously explained).

Videodisc Format

Outer diameter 12″ record	301.6 mm
Outer diameter 8″ record	200 mm
Nominal rotation speed NTSC	1800 rpm
Sense of rotation of disc seen from objective side	Counterclockwise
Thickness record and clamping area	1.1 + 0.1 mm or 0.2 ± 0.1 mm
Shape center-hole	cylindrical
Diameter center-hole	35 mm 0
Refractive index	≈ 1.5
Track pitch mean value	1.6 μ m
Minimum diameter lead-in tracks	108 mm 0
Minimum starting diameter program area	110 mm 0
Maximum diameter program area 12″ record	290 mm
12″ flexible record	280 mm
Maximum diameter program area 8″ record	192 mm
8″ flexible record	180 mm
Minimum number of stop tracks	600

Static deflection	max. 3 mm
Maximum vertical acceleration for focusing at nominal speed	≈ 10 g
Maximum radial acceleration for tracking	≈ 2 g
Maximum thickness reflective and protective layer	Within thickness tolerance of record
Maximum diameter clamping area at both sides of record	55 mm 0
Minimum diameter label at reflection side	55 mm 0
Thickness label at reflection side	≤ 25 μ m

Some Comparisons Among Equipment Types

Typical Operating Features	*Consumer Player*	*Educational/Industrial Player*
Freeze-frame	Manual selection	Manual or automatic selection
Frame identification on screen	Manual selection	Manual or automatic selection
Digital frame random access	Visual identification—manual selection	Visual keyboard identification—manual or automatic selection
Program start	Manual	Manual
Chapter selection	Automatic (manual shift from chapter to chapter)	Automatic (manual or automatic shift from chapter to chapter)
Frame by frame viewing— choice of speed	Manual selection Manual selection	Manual or automatic selection Manual or automatic selection
Forward or reverse motion (30′/sec. or fast)	Manual selection	Manual or automatic selection
Two discrete sound channels	Manual selection	Manual or automatic selection
Sound cut-off for all modes other than 30′/sec.	Automatic	Automatic
Remote control interface	No	Yes
Signal mixer	No	Yes
Standard computer interface	No	Yes
Options		
Remote control unit	No	Yes
Interactive keyboard	No	Yes
Microprocessor	No	Yes
Buffer storage modules	No	Yes
Character generation	No	Yes

Comparisons Among Media

Medium	*Advantages*	*Disadvantages*
Print	Random access In-depth study Inexpensive Readily available Replicable	Absence of motion Absence of sound Cost of color reproduction

Cinema/ Television	Audio Motion Relatively inexpensive Available Film and tape recording ability	Serial presentation/ no random access No freeze-frame
Videodisc	Random access Freeze-frame for in-depth study Inexpensive prerecorded programming Eventually available in record shops Mass replication Two discrete audio channels stereo sound dual narrative two languages commentary explication Speed regulation—slow motion forward and reverse Fast forward Chapter stop codes Automatic frame stops Manual single frame control—forward or reverse	No sound on freeze-frame

9 • *Prerecorded Tapes*

Prerecorded Video Tapes

Prerecorded video cassettes are available through large audio and video hardware stores. Also, through mail-order companies, you can order current films, old classics, cartoons and R- and X-rated adult films, documentaries, westerns, science fiction movies, and old-time silent comedies with a musical track. Prices vary from about $40 to $100. The films are available on both Beta and VHS formats. The Magnetic Video Corporation is currently selling 50 of the best 20th Century Fox films (for $49.95) as well as 8 award-winning Time-Life and BBC TV documentaries (for $35.00).

Columbia Pictures has recently set up a new division for video cassette services for duplication in the U-Matic, Beta, and WHS formats. They will also be offering their library of select films for the home market.

The films that 20th Century Fox and the other major studios are selling have all been on TV, and you may ask why buy it if you can see it on TV? One answer might be that when it appears on broadcast TV, it is usually cut for nudity, violence, or profanity. But that is not the major problem in recording films off broadcast TV. The TV networks sometimes botch up a film so far as running time is concerned. If the time slot allows 2 hours for a program, say, and the film runs 2 hours, they will usually cut this time to allow for sponsor's messages. So you would be missing some portion of the film, which (says the station) is not important to the continuity of the story.

Such problems do not arise with the many assortments of

legitimate prerecorded video tapes that are currently on the market, such as the silent films of Chaplin and the Keystone Cops. From the vaults of 20th Century Fox film library come *Cleopatra, The Detective, The Boston Strangler,* and the breathtaking *Blue Max.*

Both United Artists and Paramount Pictures have signed agreements with other corporations to distribute videocassettes of major films to home markets.

United Artists has agreed with Video Corp. of America to rent 20 major features to them. They have revealed that the rental route was one way to overcome the piracy of cassettes. These cassettes will have encoded signals on them which scramble the picture and will not permit copying.

Paramount films will go on the video cassettes' market under a license agreement signed with Fotomat Corp. Among the films for the home video market are *Godfather I & II, Saturday Night Fever,* and other such hits. Illegal stealing of Paramount cassettes will also be impossible.

For music-minded people, video tapes like *Magical Mystery Tour* with the Beatles and concerts with their favorite rock stars are also available through the companies listed in the Appendix.

Viewing Films

One advantage of the video recorder is that a film may be viewed whenever the urge arises in the viewer. Generally, the reactions of a moviegoer to a film are quite predictable: he or she follows the plot; once that is known, there seems to be little reason to view the film again. Of course, there may be exceptions, as when a particularly appealing song or dance appears, or some extraordinary special effect would demand a second viewing. To view a film only once defeats the advantages of a video recorder, where repeated accessibility to a recorded image is the prime virtue.

The question may be asked: How can we look at a film more than once and not be bored? To many movie-viewers, two or three viewings of a film is more than enough; by then many

Scenes from familiar films available on pre-recorded tapes. (1) Charlie Chaplin—*The Gold Rush,* (2) the Keystone Cops *(United Artists Corp.),* (3) Frank Sinatra—*The Detective (20th Century-Fox),* (4) Elizabeth Taylor—*Cleopatra,* (5) Tony Curtis and Henry Fonda—*The Boston Strangler (20th Century-Fox),* (6) *The Blue Max (20th Century-Fox),* (7) The Beatles—*Magical Mystery Tour*

may have memorized the dialogue and know every plot development, however obscure. But the plot of a film is not the film itself. In other words, more than the "story" is what makes a film interesting (or dull, for that matter).

On the first viewing of a film, one should relax and passively receive all the visual and audio signals sent to the brain, and just experience the film. Of course this is somewhat difficult as critical judgments are made almost unconsciously all the time. As you watch a film, the mind may comment: The scenery is pretty, the furniture is tacky, the girl is pretty, the guy is obnoxious. When we view a film we usually view it through a mental filter of our sensibilities and prejudices. Only children view films with a degree of impartiality. We all have particularly favored films–a Roy Rogers western, a Disney cartoon feature, a Bowery Boys adventure–that we saw as a child and liked for whatever reason, but *least* of all for its high "artistic" standards. The film was merely experienced and enjoyed; this is what primary viewing of a film should be.

Once the overall structure of the film is known–the plot, the scenery, the dialogue–one can begin a more sophisticated viewing, one that deemphasizes the whole and concentrates on particulars. One can take more notice of the acting (or lack of it), how the director constructed a scene, how and why the director used a certain camera viewpoint, the duration and rhythm of the editing, and so on. Every scene in a film is like a piece in a jigsaw puzzle. It can be interesting to see how they're all related.

This secondary viewing of a film can and should include "photography," which usually means composition, or how each frame of the film is composed. Almost everyone is aware when a film has "good photography." But very few people are aware of whether it is appropriate to the film's mood or intention. Good photography may mean merely picture postcard scenery and magazine cover faces. Or the photography may be an integral part of the film. Some films may have exceptional composition (almost any film by von Sternberg, for instance), others may be almost completely lacking in artful composition, either by choice or from lack of talent and knowledge.

When a viewer picks out a particular phase of a film to watch for acting, camera work, set designs, etc., the film takes on a slightly different look; it almost becomes a new film you're watching. There is so much visual and aural information pouring out of most films that one viewing cannot possibly interpret or acknowledge it all. This selective viewing of a film greatly increases one's enjoyment and understanding of that film.

There are several aspects to look for in a film to increase one's enjoyment of multiple viewings:

1. *Plot.* The basic narrative flow of the film: Who does what to whom and where and why?
2. *Acting.* Is it good, bad, or merely adequate, natural or stylized?
3. *Camera work.* Is it fanciful, artistic, or simply functional? Does the camera move, how often, and why?
4. *Lighting.* How is the film lit—are all scenes given full, flat lighting (like all TV shows) or is there atmospheric lighting, shadows, and shadings? Does the film look "dark" or "bright"?
5. *Set Design.* What is in the background and how does it reflect the characters of the actors emoting in front of it?
6. *Music.* How does the music add or subtract from the visual image? A film with too many "popular" tunes in it can detract from the story (especially if the film is not a musical). Or, does the music add to the substance of the film as does the music in Hitchcock's *Psycho?*

If you aspire to something a little more challenging, you can view a film as a sociological document: What do the attitudes of the actors, the director, and the writer reflect? A film made in the 1940s differs sharply from a film made in the 1970s. Finding out the differences can be very entertaining and instructive. The attitudes toward the government, law and order, man-woman relationships, sex, and humor have changed and can be seen by a sharp-eyed viewer.

Or a series of films can be viewed as a whole: for instance, the work of a particular director or actor. The films of Alfred Hitchcock, say, might be viewed, and common themes and similar situations observed, and his professional maturity charted. You may view the films of a particular actor or actress, for instance, John Wayne or Marlene Dietrich, and see how each has changed over the years and from film to film, physically and professionally.

Another possibility is to view several versions of the same film, say, *The Hunchback of Notre Dame.* You can compare Lon Chaney's performance with Charles Laughton's and with Anthony Quinn's; each portrayed the hunchback in a different version. The ways of looking at films are endless, as long as one remembers that there is more to a film than merely the plot.

Some films become old friends with whom we visit and of whom we never tire. How many times have we seen *King Kong* or some favorite western, musical, or cartoon, and each time been entertained and delighted? Very few films are made for only one viewing, and the best ones are always worth a repeat performance. Like love, films are better the second time around.

Piracy

Since the introduction of home video cassette recorders, the unauthorized copying of feature films has been steadily on the rise in this country and abroad. The Motion Picture Association of America spends about a million dollars a year combating film and video piracy. According to the FBI, pirates skim approximately $100 million from industry profits; they document this high figure by determining how many people would have otherwise paid money to see a film in the theater. Film pirates used to duplicate film to film, a long and expensive process. Now, with the new technology, piracy has been made very simple.

Some VTR dealers have been giving away pirated cassettes of *Star Wars* to boost hardware sales. Since many movies are shown on pay TV in the United States before theatrical release in Europe and Africa, there are large amounts of transatlantic sales. Hotel-room movies have provided pirates with a new source of income. It's not uncommon to see people checking into a hotel with a suitcase in one hand and a Betamax in the other. The FBI, with 59 regional offices, conducted over 600 raids on suspected audio and video tape piracy in the last 4 years, and seized over 4,000 video cassettes and films. Nevertheless, today, if you have the contact and the money, you can buy a pirated copy of any film in release, from Chaplin to *Superman.*

Copyright Law

Since last year, there has been an increasing number of FBI raids, arrests, seizures, indictments, and convictions of manufacturers, distributors, and importers of bootleg video tapes.

In Portland, two defendants were found guilty by a jury on eight counts of copyright infringement and conspiracy. The two defendants had been charged with the duplication and distribution of pirated tapes. For this trial, an actual tape duplicate was set up in the courtroom to demonstrate graphically to the jury how it is done.

The state of Wyoming became the 49th state to enact an antipiracy law outlawing the manufacture and sale of pirated and bootlegged audio and video tapes. Vermont is the only state still remaining without a state antipiracy statute. The Wyoming law makes it a felony, punishable by two years' imprisonment and a $10,000 fine, for the unauthorized duplication of sound and video recording.

Kentucky has upgraded its laws on piracy. The crime of manufacturing pirated tapes has changed from a misdemeanor to a felony, punishable by not less than one or more than five years' imprisonment and/or a fine of $500 to $3,000. Kentucky's new law has a provision requiring peace officers to seize any such recordings, and recordings so seized are required to be forfeited and destroyed by enforcement authorities. In January of 1978 several of the movie studios took out an ad in *Variety,* a weekly trade newspaper for the entertainment industry, directed to the buyers and collectors of pirated tapes and films. The message was quite simple.

NOTICE

The Copyright Act of 1909, as amended, confers upon the owner of a copyright the exclusive rights, among others, to copy, vend, and publicly perform the copyrighted work.

The new Copyright Act (Public Law 94-553) which becomes effective on January 1, 1978, also confers upon the owner of a copyright the exclusive rights, among others, to reproduce, sell, and publicly perform the copyrighted work.

Any infringement of these and the other exclusive rights of the copyright owner gives rise under both the Copyright Act of 1909 and the new Copyright Act to a cause of action for the seizure and impoundment of infringing prints, and for injunctive relief, damages, and attorneys' fees.

Moreover, under the new copyright law, which will become effective on January 1, 1978, Congress has expressly disavowed the Foreman decision and has clearly stated its intent that "the burden of proving whether a particular copy was lawfully made or acquired should rest on the defendant"–House Report No. 94-1476, page 81. In other words, it is not enough for you to just demonstrate the source from whom you acquired the prints and that you paid for them, but you will be held responsible for showing that the seller (and his seller) had the right to sell the particular prints.

In other words, the studios are serving a warning to collectors of bootleg art that they were not going to ignore video tape collectors who fill their libraries illegally the same way film collectors did.

10 • *The Camera*

Next to the Kodak Pocket Instamatic, a TV camera is the easiest camera to operate, mainly because of the electric viewfinder, which is a small picture tube, sort of a miniature TV set. This eliminates the fuss and guesswork of light meters, and depth-of-field focusing.

The heart of the video camera is a tube called the vidicon, which contains a screen upon which light is focused. Other types of camera tubes with greater sensitivity are available, called the tivicon or silicon diode tube. These cameras can record an acceptable picture in extremely low light. The lenses in a video camera are just like the lenses on any camera with a C-mount. There are four different types of lenses:

1. *Wide-Angle Lens.* This lens gives a very broad picture of things close to the lens; also, everything seems farther away than it is.
2. *Normal lens.* This is equivalent to what the eye can see.
3. *Telephoto lens.* This lens will give a narrow picture of things far from the camera.
4. *Zoom lens.* The zoom lens (wide angle to telephoto) is many lenses combined in one. To focus the zoom lens, it must be zoomed into the telephoto position. The focusing is done by means of the focus ring, which is at the end of the lens barrel. Once in focus, the lens will remain so as it zooms back to wide angle.

Compatibility is not a problem. Any video camera can be used with any video cassette recorder. Some of them are

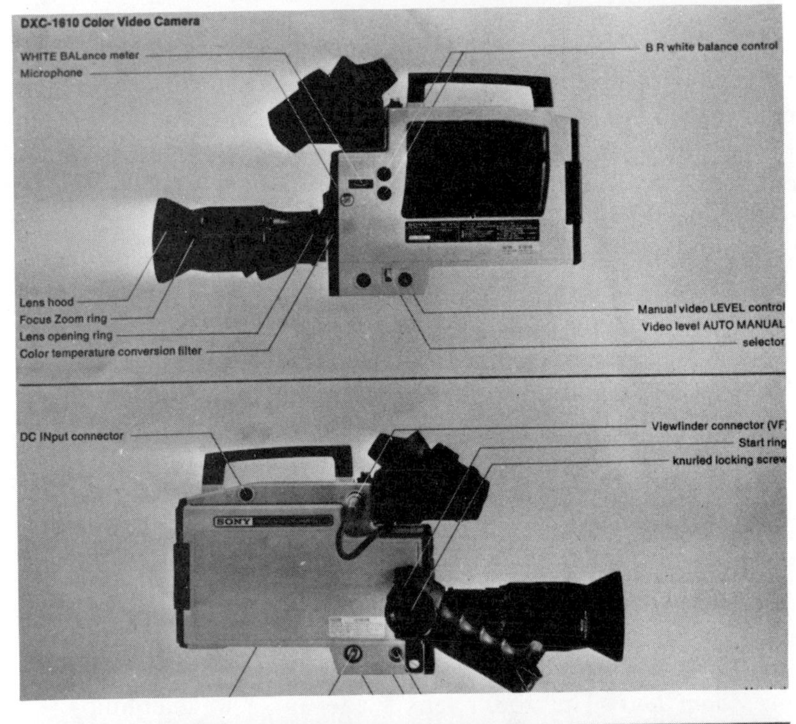
DXC-1610 Color Video Camera
WHITE BALance meter
Microphone
B R white balance control
Lens hood
Focus Zoom ring
Lens opening ring
Color temperature conversion filter
Manual video LEVEL control
Video level AUTO MANUAL
selector
DC INput connector
Viewfinder connector (VF)
Start ring
knurled locking screw
SONY

SONY

Various types of camera: (1) The GBC-VF 302 camera with 22mm to 90mm zoom lens; (2 and 3) The Sony DXC-1610 color video camera; (4) The Sony AVC-3400 model; (5) The Sony AVC-3450 camera.

equipped with circuitry that automatically adjusts exposures to available light conditions.

And how does this magnetic video recording process work? The camera perceives a scene and converts it into a series of electric signals, which are sent to the video tape recorder. The VTR processes these signals and sends them to the video recording head. The function of the recording head is to create a magnetic force that acts upon the passing tape with varying frequency and intensity. The magnetic force generated by the record head magnetizes the magnetic oxide particles in the tape, creating a magnetic pattern. These magnetic patterns form a memory of the original scene on the video tape.

On playback, the video head senses the magnetic patterns on the tape, and generates electrical signals which pass through the system and into a monitor or a TV set.

Camera Notes

It is important never to shoot directly at or accidentally point at a bright light. This action could cause a dark spot or several spots on all the pictures called a "burn." If the burn is not too serious it can be erased by aiming the camera at a neutral background out of focus. A very serious burn can destroy a vidicon tube.

The vidicon tube, although sturdy and efficient, operates best with a good light source. Improper lighting conditions will cause the picture to appear gray and grainy.

CLOSED-CIRCUIT VIDEO SYSTEM

11 • *Optional Equipment*

Video Editing

Video editing is basically a process of dubbing: You take a piece of information from one tape and transfer the section onto another piece of blank tape by a set of coaxial cables.

Editing is basically a matter of assembling pieces of tape A, a bit of tape B, and a bit from tape C, and so on, onto a final tape. It is true that copying one tape onto another tape does result in a slight loss of quality.

You must use two video decks that have a common sync pulse; however, there will be considerable breakup (noise) in the edited picture at the beginning of each new segment.

One big problem is avoided by using a machine that has specific editing capability.

Another problem is that you may not like the finished edit. You simply back up the machine and come in with a new shot a little earlier. But if you cut the scene too soon it means that the entire edit of the previous shots must be repeated.

There are also problems with sound; the head that reads the sound may lag behind the head that reads the video by a half a second, causing audio lag. This means that when an edit is made onto a picture cue, it takes a half a second for the sound track to catch up; the previous sound will play with the new picture for half a second.

There is usually a "pop" when the audio dubbing button is pushed because of the immediacy of the electrical contact. Reciting the problems with small video format editing could go on and on.

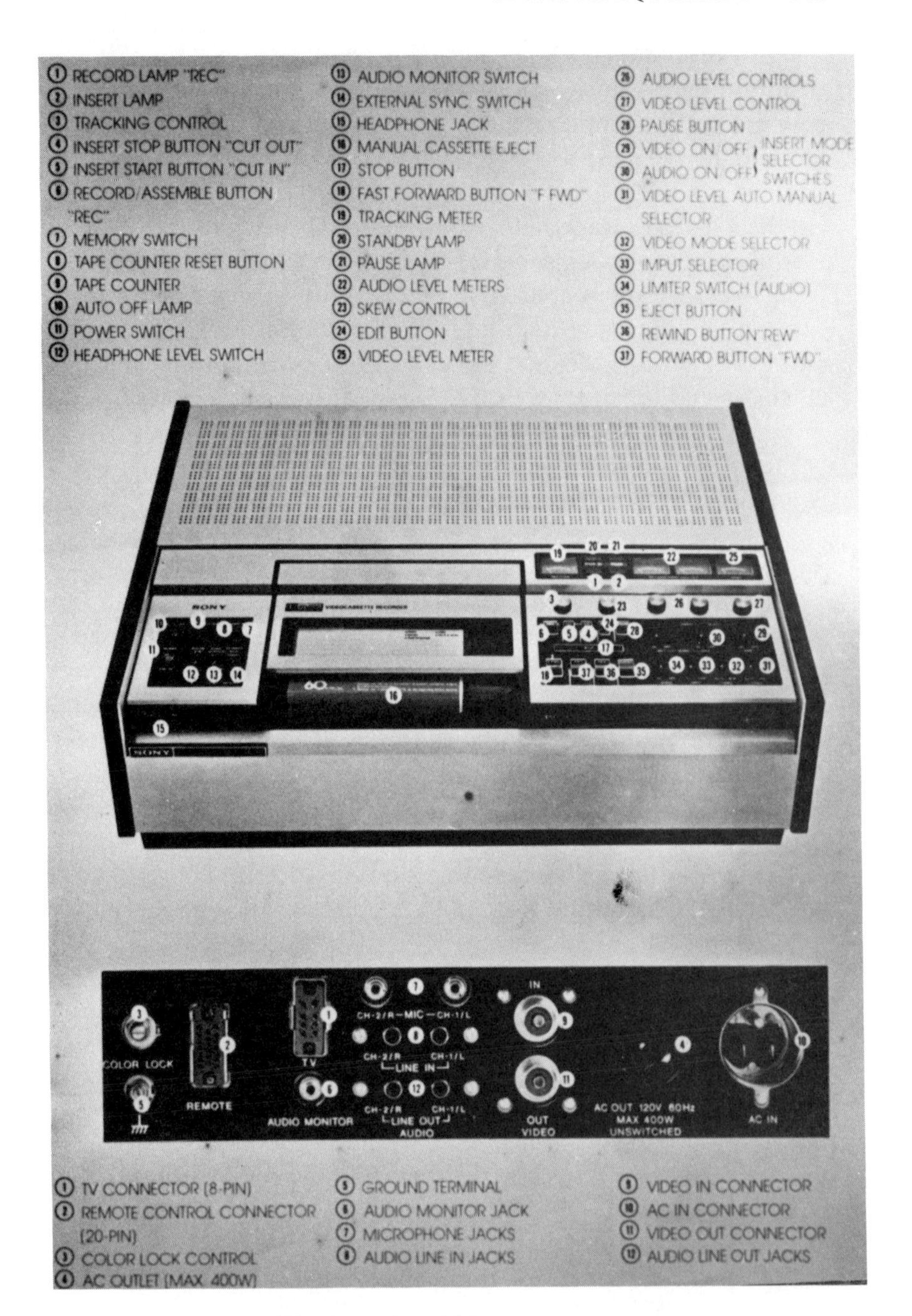

The Sony VO-2850 editing deck

But four years ago Sony started a revolution in tape editing by introducing the VO-2850, a three-quarter-inch U-matic tape editor. The all-around results of editing on the VO-2850 have to be seen to be believed. The switch from playback to record mode is made during the vertical blanking. Rotary erase heads erase each video track before new material is recorded.

It also incorporates such features as these: slow-speed playback; freeze-frame capability; precise edit point; separate recording facilities for video and the two sound tracks; a search mode that permits location of cueing at one-fifth normal speed; glitch-free edits; and the lack of "pop" in the audio track.

Video editing falls into three main modes, and is done with two machines that are interlocked.

Assembly Editing

This consists in recording a portion of video onto the end of video already on the main program tape. When this new section is put into the tape, the material in front of it is never returned to, except in playback. Once a new section has been added, further sections can be similarly added to make a complete program.

Insert Editing

This procedure places a new section of video into a program, and at the end of the new material the original program is returned to.

Audio Dubbing

The ability to record audio only onto the tape while the video circuit remains in the playback mode is meant here.

Assemble Editing from Another Video Tape Machine

Two machines are hooked into each other. One machine holds a recorded video program; the second machine will hold a playback tape.

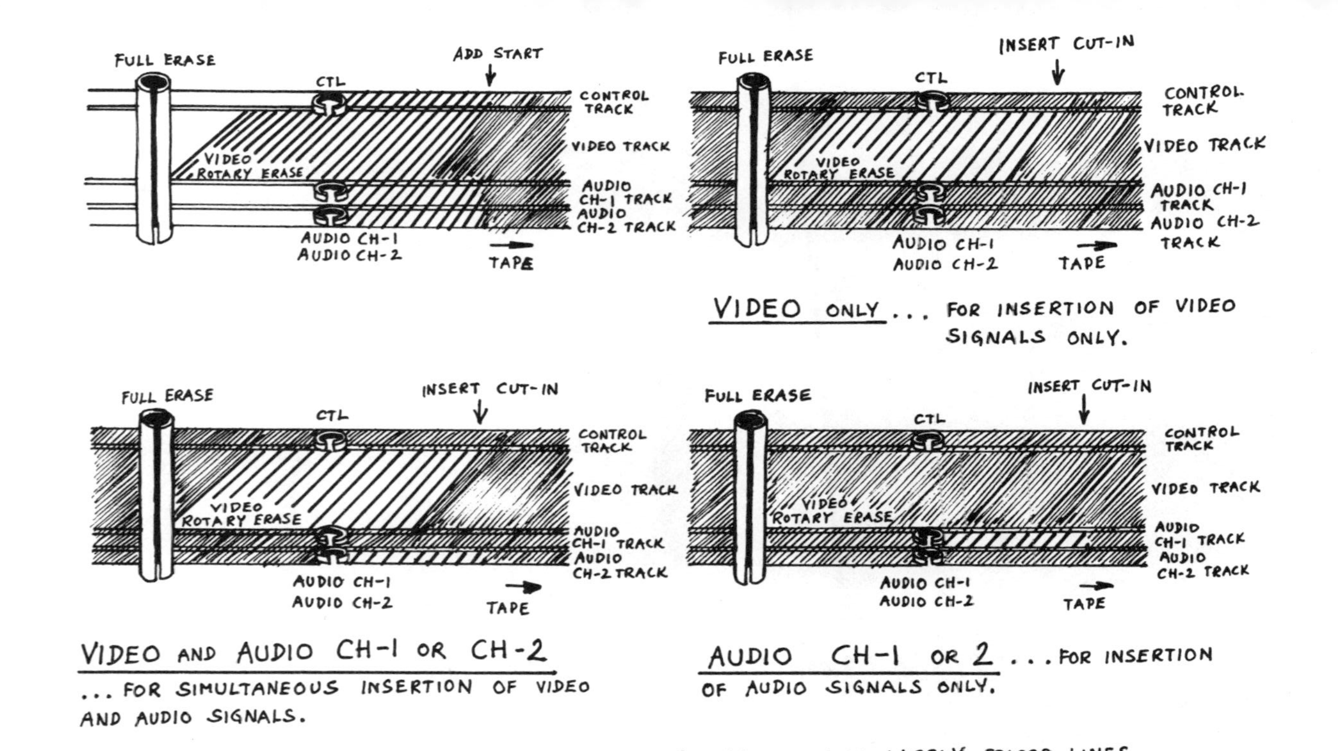

(left) Assemble editing: adds new video and audio signals onto a prerecorded tape. (right) Insert editing: Inserts new materials onto a prerecorded tape and has three insert functions.

The concord monitor and GBC camera

The section of the playback tape which represents the start of the program material to be added to the main program is decided, and this is marked cue 1. A point about 15 seconds before this is determined and is labeled cue 2. The entry position for the new video is found on the main program, and this is called cue 3. It is now necessary to decide cue 4. This is the position in the main program where the main tape will be left stationary, and this position is about 10 seconds before cue 3. The main tape is set up at cue 4 and left.

The playback tape is now started from a position about 35 seconds before cuc 1. When cue 2 appears, thc main program tape is started. The tape now locks up to the video incoming from the playback tape, and if cue 4 has been correctly chosen, then cue 1 and 3 will meet at the same point. If they meet, then the tapes are reset, the procedure is repeated, and the edit cut is finally made.

Insert Editing from Another Tape

Insert editing is more difficult than assembly editing. The tape is played up to the point where the program is to start, and the cut-in or edit button is pressed. The editor must now time the action with a stopwatch and must finish at the correct time so that the cut-out button can be pressed before the resumption of the original required video.

Audio Dub

Audio dubbing can be done from any source, with either the mick or aux inputs which can be used separately or together.

Monitor

Although a video cassette recorder can be wired to any normal TV set by means of an RF adapter, the cleanest clearest pictures in playback will come from a monitor. A monitor is a TV set that receives the tape signal directly, bypassing the tuner. The monitor/receiver can receive regular TV programs, so it is possible to record television programs off the air.

Time-Lapse Video Recording

Today's video technology has brought us to the point where it is possible to record 108 hours in real time in a single-hour roll of half-inch tape. How is it possible to fit 108 hours of time into 1 hour of tape?

Obviously, the farther a tape travels past a stationary recording head, the more information can be stored on the tape. This applies to audio recording as well. Helical-scan video recording achieves the necessary tape-to-record-head speed ratio by moving forth the tape and the recording head simultaneously. In helical recording, the video track lies at a 10-degree diagonal to the length of the tape. One TV field or frame is written each time the head crosses the tape. To produce a time lapse effect with a helical VTR is to record every tenth field of incoming signals.

Security Systems with Time-Lapse Video Recording

Alarm sensors, such as door, window, or mat switches, are connected directly to the VTR alarm input. When one of these switches is closed, the VTR goes into the alarm mode. A front panel alarm light goes on, and the alarm output allows external devices such as warning light and buzzers to be used.

After approximately 5 minutes, the VTR will then resume its original recording speed automatically. It can also be externally or manually reset. Even though the VTR will again be recording at its original speed, the alarm indicator and alarm output remain on. This is important. Suppose an alarm is tripped while the security guard is making his rounds. He later returns to the VTR to find the alarm light on. He will know that at some time in the past an alarm had been set off.

Until the tape has been searched completely, however, the guard will not know when or why the alarm was set off. Searching the tape conventionally could take hours, even days—at original speed! But the unique alarm memory feature makes it possible for the tape to be searched in a matter of minutes.

How is this rapid tape search made possible? Half-inch video tape recorders display a pattern of scan lines and other video information when the tape is moved at high speed during rewind and fast forward. When the VTR is put into rewind or fast forward, the monitor screen goes completely blank and the alarm lamp lights when it reaches the portion that was recorded during the alarm period. Once that portion is passed, the familiar pattern of video information is again displayed. When the VTR is put into play, its alarm light goes on at the point on the tape when the alarm had been set off and stays on during the alarm interval. Users will appreciate the convenience of high-speed playback. When a tape is recorded at any low speed, it can be played back at any of the higher speeds without disturbing the horizontal lines in the picture. Thus many hours of material can be quickly scanned.

To insure the best possible video signal on playback, a front panel switch adds a vertical sync pulse to the output signal. In certain cases, this will improve the stability of the playback picture.

The maximum time lapse recording time is 108 hours, but up to three VTRs can be connected together to provide a maximum continuous recording time of 324 hours—almost 2 weeks, recording 24 hours per day! An AC power outlet at the back of the VTR turns on at the end of the tape to start the next VTR. The AC outlet can also be used to turn on a buzzer or warning light to signal the end of tape.

Super-8 Film-to-Video Tape

Videoplayer VP-1

The Kodak Supermatic Film Videoplayer is used to feed Super-8 film images to a single television set or to an entire TV network. With the VP-1 Videoplayer, companies using both individual film projectors with closed-circuit television for sales and training can produce and distribute all programs in one medium.

Videoplayer VP-X

The Kodak Supermatic Film Videoplayer VP-X offers a practical and economical means to broadcast Super-8 film—color or black-and-white, sound or silent. Super-8 films from amateur photographers shot at 18 fps or films shot with the new single-system sound cameras at 24 fps can be broadcast direct because the videoplayer serves as the film chain: no more need to transfer to tape or blow up to 16 mm. Place your Super-8 film on the videoplayer. Push a button, and the film threads automatically.

In addition to film frame rate, there are also controls for still frame, as well as for framing adjustment for running and still, steadiness adjustment, blue/red balance, and focus. The Videoplayer automatically rewinds and returns to standby at the end of each film.

The Kodak Supermatic Film Videoplayer VP-X accepts external synchronization signals to lock the videoplayer to station sync. A VP-X has the same 1-volt peak-to-peak composite video and preamplified auto outputs as the VP-1, and requires the following sync signals: burst flag, composite blanking, vertical drive, composite synchronization, horizontal drive, and 3.58 color subcarrier.

A VP-X brings a new dimension to news and other programming, and gets you directly on the air with virtually any piece of Super-8 film—ideal for pickup of unique newsfilm footage captured by amateur photographers at the scene: also, for school and community events and programs produced on Super-8.

A VP-X also offers an extremely economical film-to-tape transfer process. With two units, you can A & B assemble quickly, cleanly, and without rollover.

When used in conjunction with servo-capstan VTR equipment, a videoplayer lets you insert and assemble Super-8 footage onto videotape. No time-base correction is required.

Super-8 film and inexpensive equipment offer the lowest cost as an aid to get into your own film origination. Professional quality news, commercials, and community-service films are feasible without complicated equipment or techniques. A simple, inexpensive means of program origination for regional or satellite broadcasters, Super-8 is also an excellent support and backup medium for reporters, cameramen, and auxiliary personnel of major stations. A VP-X also fits well into other video operations requiring systemwide synchronization, such as those found in the business, industrial, and educational fields.

With economical Super-8 film, color is within reach of even the most modest budgets. A VP-X lets you broadcast color with a minimum investment in equipment.

A VP-X offers particular potential for cablecasting. It not only gets you directly on system with Super-8 film, but when compared with the substantially higher cost of alternative film-chain equipment, it provides an outstanding image on the television screen.

Portapak

If you are planning to be using your portapak on a special assignment, you must ask yourself these important questions.

- Will electrical power be available?
- Will there be adequate light?
- Should a special long-life battery be used?
- What kind of microphone should be used?

Once you find the answers to these questions, you can proceed to pack your own equipment. The packing should consist of the following equipment:

1. Portapak itself
2. AC power adapter. This AC adapter enables the camera and deck to work on a normal house current and also acts as the battery charger.
3. Video tapes
4. Extra batteries
5. Microphones

6. Headphones
7. Extension cable
8. RF modulator
9. Lights

Microphones

When working in a studio, or located with a portapak, recording excellent sound as well as video is very important. Selecting the right microphone for your needs will help to give the look and sound of a polished production. You must remember that in-camera mikes that are electret condenser omnidirectional microphones are good for the person who is shot by himself and wants to capture general sound.

We recognize four microphone types:

1. *Crystal or ceramic microphones* are crystal or ceramic elements to produce the voltage. They are excellent for certain transmitters, like walkie-talkies, but are not meant for audio used in video production.

2. *Dynamic microphones* have a small wire coil that moves back and forth around a circular magnet to produce the voltage. The wire coil is attached to a small diaphragm that is vibrated by the sound waves, causing a signal to come out of the microphone.

3. *Condenser microphones* are high-quality mikes that absorb a wide range of sound effectively and have a higher signal output. Condenser microphones are delicate and sensitive to temperature and humidity. All recording studios employ the condenser microphones, which are also relatively expensive and require an external power supply.

4. *Electret condenser microphones* have as their active element an electric circuit that acts as a capacitor. The capacitor retains a charge that can be run off very small batteries inside the shell of the mike. The electret condenser microphones have a wide frequency response.

A disadvantage of electret condensers is that they are not as durable as dynamic mikes and will damage easily if dropped.

Also, their batteries may burn out at a critical time when recording. It is impossible to ascertain the condition of the batteries inside the mike, so it is necessary to carry an extra set around.

Another way to distinguish microphones is by format, a term that refers here to their range and the direction from which they are able to pick up sound. Again we have four classifications.

1. *Omnidirectional.* Microphones pick up sound coming from all directions, although they favor the direction in which they are pointed.

2. *Unidirectional Microphones.* These pick up sound primarily from the front and only a little from the sides. The frequency response is not as flat as that of the omnidirectional mike.

3. *Lavaliere Microphones.* These are small mikes that are worn around the neck or pin on a lapel. They are very good for picking up only the voice of the wearer.

4. *Shotgun Microphones.* These consist basically of a long tube containing acoustical resonating chambers connected to the microphone. These chambers pick up sound coming from the front and cancel sound coming from the sides. The thing to remember when using a shotgun mike is that they are far more directional at higher frequencies and will tend to pick up some lower, bassier sound coming from all directions.

12 • *Film Production and Other Uses of VTR*

Videotape to Film

Before video tape recording started, there was only one way to record a program produced for television broadcast, a system called Kinescope. The Kinescope films the face of a television tube during the actual broadcast. The Kinescope process is the most widely used of the methods for the transfer of video information to film.

In 1963 Richard Burton played Hamlet before the video camera. It was the first time a movie was shot in videotape then transferred to film for major distribution in the theatres. Five other productions followed suit; *Harlow* (1965), *The Tami Show* (1965), *The Big TNT Show* (1965), *The Committee* (1969), and Frank Zappa's color masterpiece, *200 Motels.*

A Tape-to-Film-to-Tape Production

Sticks and Bones was to be aired on March 19, 1973, but was cancelled by CBS for being out of step with the temper of the times. *Sticks and Bones* was shot in video with the use of only one camera. Video was financially feasible and proved to be less energy-consuming for the director. He didn't have to demonstrate to the actors how a scene should be played. After shooting a scene, the tape could be viewed by the actors and director. The actors could see for themselves what they did wrong and go back and correct it. With this system the actors did all the work for themselves.

The playback capacity of tape gives the film maker a complete copy of everything, and the fear of the lab messing things up is eliminated. Other advantages to this system are that it is quick, as well as being technically simple to operate, with the capabilities of producing an instant replay. Instant replay is extremely important when viewing a scene because of its lags; the film maker can go back and re-shoot, thereby saving a tremendous amount of time. Unfortunately, editing tape is very tricky and expensive, costing somewhere between $200 to $300 an hour.

So the film makers shot the tape on 16 mm, edited it, and then reshot it back onto tape. This made *Sticks and Bones* a tape-to-film-to-tape production.

Film Production

Video may be the means that producers and directors have been looking for to show and explain proposed ideas for filming. Up until this time a storyboard or sketch pad were used to give a prospective client an illustration of what would take place on screen. This often resulted in a client's complaints or rejection once the film had been shot. Video taping costs are nominal, and a client can get an idea of what the final product will look like before a lot of money has been spent on film production. The videotaping is done with simple lighting and stand-in talent and shows clearly what the spot would look like. Video equipment can be rented inexpensively, but it is advisable for film production companies to purchase video equipment because of its obvious other uses for film producers. An example of this would be using video to test talent instead of showing photographs, a means that can be stored for future reference.

It can also be used to scout locations. Previously some companies used Super-8, but the footage was shot in short segments. Video, on the other hand, can be shot in long segments and is cheaper to use than the Super-8 equipment.

A video camera can be used during the shooting of a film to check for continuity in a wide-angle shot, as well as to avoid costly mistakes by showing, in terms of highlights, how the scene's lighting will appear. Once video is acknowledged in the film business it will prove itself to be invaluable as a preproduction tool as well as a valuable production tool that can help generate the film business.

Storyboard

The most important step in any production is called preproduction. This term means the careful planning of what you're going to shoot and how you're going to shoot it. This is when you employ a storyboard, which is a series of drawings depicting your story scene by scene. It is always better to put everything on paper before you pick up a camera to start your production.

Professional people know that the more preplanning you do prior to the actual shooting, the fewer surprises. Storyboard scripts should be written in two columns, one for video and the other for audio. Camera movements must be written into the narrative and used to focus the viewer's attention. The pictures should move meaningfully from place to place.

Lighting for a Video Production

The Three Main Lights

1. *Key Light* (the main source of illumination for the subject in the picture). Use a lamp with a high-wattage bulb and place it a short distance to the right or the left of the video camera, fairly close to the subject. The illumination of the subject will be strong on one side and slightly dark on the other side.
2. *Fill light.* Less intense and direct and intended to provide general illumination.
3. *Back Light.* Back light provides a barely distinct rim of light on the subject's head and shoulders and serves to distinguish him from the background.

News Media: Journalism

Since the convention of 1968, electronic news gathering has been making headlines. The last presidential nominating conventions showed the latest electronic equipment at work. The first portable camera, the big Ampex VR 3000 quadruplex, weighed 45 pounds and could be held for only a half hour at a time–anything but portable. In 1972 the Ampex was replaced by the 32-pound Ikegami. This camera made shooting a little easier and is still in use today.

The Time Base Correction

The time base corrector is a machine which will accept a recorded tape, to be played back with all of its time base errors, and will remove all those errors, producing a stable picture with a rigid sync signal and so making it usable for broadcasting systems. The introduction of the time base corrector allowed stations to use the U-Matic video tape recorders for broadcasting.

At the last political conventions the RCA TK-76, weighing 19 pounds, was used. It plugs into a 12-volt battery belt that weighs about 6 pounds. The camera can also operate on a DC, 12-volt cigarette lighter. It features an adjustable viewfinder and costs $35,000.

CBS Labs developed the Microcam camera, weighing 8 pounds. The CCU is worn around the waist and carried over the shoulder and includes a rechargable battery that runs for one hour. It can be made to run up to 8 hours by using a standard 12-volt battery belt. Unlike other video cameras, it can work in low-light levels of 5 foot-candles, using a 1.4 lens opening. Lighter than many 16 mm film cameras, it uses 23 watts of power. The Microcam is a wonder at $30,000.

Both the RCA and Microcam can either record or give a direct live transmission. The cameras are either hooked up by a cable system or use a microwave transmission, which cuts out the cost of cabling large areas to a U-Matic recorder. Because of these advances in equipment, station wagons and small vans

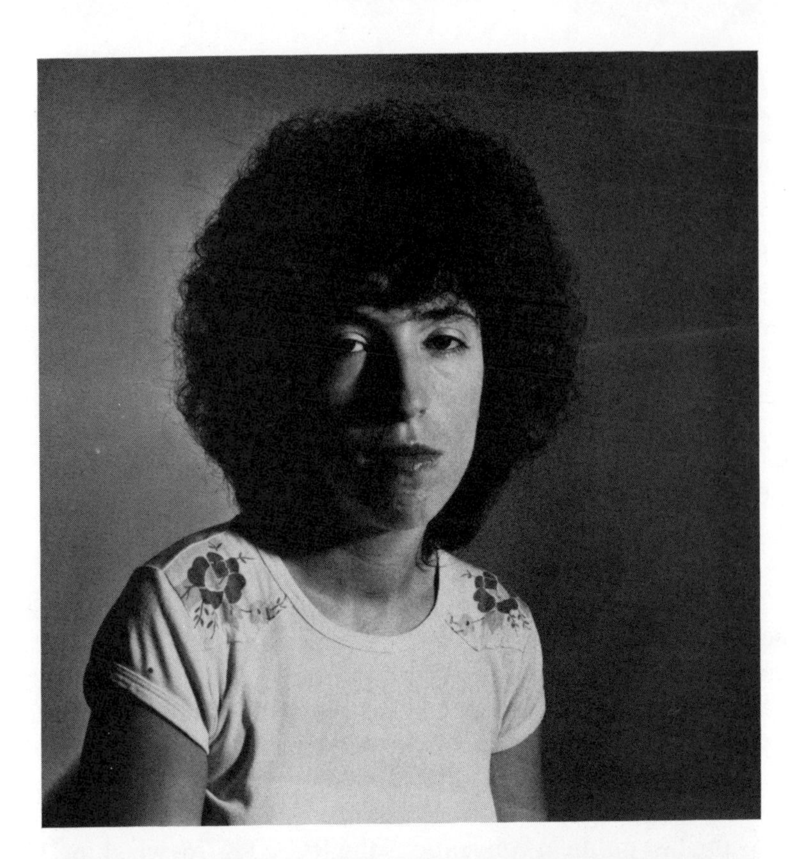

(a) The lighting effect from using three lighting units—main light (key light), fill light, and back light (rim light); (b) the result of taking away the main light and just using the fill light and back light; (c) Using just the back light.

are replacing the large mobile units of the past. The taped cassette is brought to the studio for editing on broadcasting.

The new BVU-200 portable recorder has a greatly improved signal-to-noise ratio and improved stability. It can also work on batteries to play back on color and still maintain an improved quality in the picture. All of these improved features in the U-Matic recorder are important to reporters because of their constant mobility.

The new BVU-200 can be hooked up with the BUE-500 editing console. There are two audio tracks and a simple address track for editing inputs. Since the BVU-200 doesn't disengage the tape from the control track head in search and preview, accurate edits are made possible. Its most valuable features are its ability to program edits, as well as the new forward and reverse fast-search function.

Editing is made easier because of the accurate frame counter, which shows frame number, minutes, and seconds. The console has four different tape speeds, which allow the operator to shift frames. The operator can also trim frames at the entry or close of an edit. The machine can be adapted to such use, with memory logic mode switches.

Criminal Justice

For many years the commercial television industry has dominated the video tape medium, never fully exploring its potential. Closed-circuit, portable video tape is probably the most important media tool of the 1970s. Recognizing the future of the medium as a powerful teaching aid, political force, and creative medium, trained individuals from such diverse fields as education, social work, business psychology, religion, and art have been experimenting with video tape as a tool to further their professions.

In the field of criminal justice, video is seen as a tool dedicated to positive social change. In 1974, New World Communications taped a show run by the inmates of Terre Haute Federal Prison. Filming was done and completed in one day of shooting and consisted of convicted felons talking with young people in small rap sessions, as well as lectures. These firsthand accounts seemed to have more effect upon kids headed for trouble with the law than conventional warnings. The finished show won state and national honors in the field of criminal justice.

The criminal justice field has started to explore the possibilities of video applications in the area of trial process. The growing interest in video applications proved to be unlimited, the only obstacle being the reluctance of judges and lawyers to accept this innovation. The traditional procedures of law are based on precedent and resist change.

The first judge to allow video equipment into his court room was Judge John B. Wilson of Marion County Criminal Court No. 4, of Indianapolis, in June, 1974. With the consent of both the defense and the prosecution, a semi-portable, black-and-white Panasonic was set up in the court room, but merely served to provide an accurate time reference. The video gave a good picture of both the lawyer and the witness during questioning as well as demonstrating that it was possible to video tape a trial without totally disrupting the normal proceedings. Once the trial was under way, those people participating in the courtroom, the judge, the prosecution, and the defense, all said they forgot about the video taping.

At subsequent trials the video tape was used as a supplement to the written transcript and later proved to give a more accurate account of a trial than the transcript. Both forms of recording were compared to clearly determine their accuracy. The transcript was revealed to have several important errors as well as causing problems in identifying who had said what.

Because of these favorable findings, video tape has inspired interest in other uses within the criminal justice system. The first prerecorded video tape trial took place in August, 1974 (State of Indiana vs. Marks), and video tape was the only record of the trial. A major difference in the trial procedure created by the use of video tape was that the trial was conducted in the absence of a judge or jury with all inadmissible evidence edited out before being shown to the jury.

The advantages of using video are quite evident. First, if a defendant fails to appear the judge and jury have to waste half a day trying to locate him. With video tape, the testimonies can be taped over a lengthy period of time without the postponement of the trial. Secondly, it removes the impressions made in a jury's mind when a lawyer intentionally offers inadmissible evidence, and an objection is made. The jury is then instructed to disregard the testimony but an impression has been made. The use of video eliminates this strategy employed by lawyers, because all objections, if overruled, are edited out, and all objections that are sustained are left in after the judge views the tapes and makes his rulings. The original tape always remains intact while the edited form is shown to the jury. To illustrate, in one case 2½ hours of testimony was reduced to 1 hour and 20 minutes after editing. Opening and final arguments were given live and, by a vote of 10 to 2, the jury favored the prerecorded video tape trial over the traditional trial.

Video can facilitate the justice system once the controversy about it has been settled. Video has demonstrated its potential as a time-saver, but it also has many other potentials not yet explored, as in the case of a retrial or the death of a key witness. If the original tape were on file it could be shown reedited to a new jury without retrying the case. The numerous instructions a judge must repeat to each jury member can be taped, as well as deposing of testimonies of expert witnesses like doctors and lab specialists, who have crowded schedules that impede court scheduling.

At present, video tape is not being eagerly accepted by judges and lawyers; and it might very well take a Supreme Court ruling before they will finally accept it.

Cable TV

Cable TV is growing, and with the increase comes the battle of different cable firms for each town. Cable companies must win territory by promoting their program package—at stake are millions of dollars in profits. The companies will stop at nothing to achieve substantial lists of subscribers. By pointing out their good points, and their opponents' bad ones, the cable conflict stops little short of libel. The companies accuse each other of misrepresentation and incompetence when a city is deciding between two companies. Town officials are often invited by the companies to inspect their studios.

Because there is no exclusive monopoly there are no laws that stop a cable company from applying for a franchise where one has already been granted. The battles and the conflict are a matter of life and death because a franchise provides a monopoly for a good 10 to 15 years.

Public TV

For the Bicentennial, Virginia Kassel produced *The Adams Chronicles,* which consisted of thirteen meticulously researched one-hour dramas, shot entirely on video tape for public television. This was the largest dramatic production ever attempted by noncommercial television. *The Adams Chronicles* was shot on location and in refurbished twin studios by the various directors who filmed the episodes. Shooting on location was extremely difficult because multiple video tape cameras were used that saved time and made two camera sequences worth all the trouble. Switched multiple cameras were used, as opposed to film-style single-camera shooting. Entire scenes were rehearsed to be shot as for television and shot in one long stretch.

Lighting proved to be a major problem for the production company. The only source of light in *The Adams Chronicles* were candles and hearth fires, which can easily be handled by film but which cause a problem for video systems. Very often the video system was working close to its limits.

Some TV directors felt single-camera film-style shooting allowed them greater creative freedom because they were unhampered by the restrictions of multiple camera shooting. By using a single camera they tried to exceed the norm of video, using techniques used by film cameramen for years but only recently introduced to television.

Editing was done under the supervision of each episode's director. Shows shot with switched multiple cameras needed less time as a whole compared with those shows shot with a single camera.

The Adams Chronicles is only one example of the possibilities of video used for television.

Commercial TV

For commercial television, advertising time represents millions of dollars and, therefore, the advertising industry devised some pretty ingenious methods of adding more commercials to prime time programming.

The television code of the National Association of Broadcasters allows 9½ minutes per hour in prime time, and 16 minutes per hour at other times for nonprogram time. Unfortunately, membership in the NAB is voluntary, and not all stations belong. Controls and limitations on commercial time result from a gentleman's agreement maintained by both the broadcasting industry and government standards. Fortunately, all stations are responsible to the FCC, which serves the public's interest. The FCC grants the licenses, and a station must state how much time it plans to devote to commercials. If it exceeds the 16 minutes per hour it must demonstrate that the commercial being shown serves the public's interest. But in reality the networks need only answer to themselves because the standards are self-imposed.

As the desire for advertising time increases, it is just a matter of time before they exceed the limits they have laid down for themselves. Unfortunately, these other seconds will come out of programming time. An example of this is the extra 30 seconds NBC has created for its affiliates to sell during *The Bionic Woman* and *Little House on the Prairie.* Needless to say, there is great controversy over this increase in commercial time.

The Spinks-Ali heavyweight championship fight is a good example of how the networks cluttered ads into a 2-minute break. Some commercials ran past the starting bell and in the event of a knockout in the first few seconds of the fight the home viewer would only have been able to see the knockdown by an instant replay. Sports has seen the highest escalation of commercial time–22 minutes of advertising per game.

The major networks broadcast their programs and commercials with a certain amount of entitled time for local network spots.

The latest gimmick to gain rapid acceptance is 10-second plugs during the 1-minute news break, which first appeared as the Bicentennial minutes. Sponsors pay as much as $30,000 for a 10-second plug, which is as much as they normally would pay for a 30-second plug. A network can fit in 12 of these 10-second plugs during a 2-minute break but are reluctant to do so because of the cluttered appearance it projects, making the idea relatively remote.

Industrial Sales

Independent film makers are finding video tape industrial films to be a dependable source of income. Because of the great portability of video equipment as well as the simplicity of cassette playback the video cassette can be seen as the ideal medium for sales.

Many industrial users favor the Sony U-Matic video cassette system, using a three-quarter-inch tape cassette and playback unit that can be plugged into any color TV set. This Sony equipment is simple enough to be operated by untrained people with just the base minimum of instruction.

Some valuable features for the use of video in sales are these: (1) The tape can be stopped at any point and reviewed. (2) Because of its portability the salesman can bring the equipment right into a customer's home. (3) A three-quarter-inch cassette weighs about 1.2 pounds, so each salesman can carry a minimum of six easily. (4) A large company with widespread branch offices could use the cassette to provide information to its personnel.

The Burroughs Corporation uses a library of video cassettes to keep its sales force up to date with their production line. This gives the salesman the advantages of being able to review a tape whenever he wants an instant refresher course.

Wilson Learning Corporation uses video cassettes to train life insurance salesmen by showing them sales situations, supplemented by workbooks. This program has been so successful that Wilson will soon be establishing their own television production facilities.

The General Telephone Company of Florida uses cassette tapes for training their staff: Previously, the company would transport the entire staff to Tampa—a very expensive procedure.

The fashion industry has many potentials for the use of video equipment. Presently Lord & Taylor uses the Sony BXC-5000B system for keeping their staff informed about new merchandise and fashion trends. The tapes are produced by the store's buyers and managers in their own studio, set up in the New York store.

Video cassettes are feasible and attractive to any number of industrial users with many opportunities for money making projects. The possibilities are countless.

Marketing

"Money News Inserts" were to be 60- and 90-second spots on the subject of money, slated to be syndicated in local news markets. Persons working in cooperation with the staff of Time-Life's *Money* magazine researched concepts, followed up with interviews on location such as the Okefenokee swamp.

The technical excellence and flexibility of the Asaca 3000 camera proved vital to a one-camera film production crew. The crew consisted of a field producer, camera operator, and VCR (video cassette recorder) audio operator—no engineer. The VO-

3800 VCR proved to be an efficient, dependable machine throughout the 16 months of filming. On the road, routine maintenance was performed daily, with a test cassette used to check service and transport functions, as well as the color balance of the camera.

The crew knew the limitations, potentials, and test procedures of the equipment and could produce superbly engineered cassette tapes. A three-quarter-inch cassette was used as a workprint for editorial and scripting purposes.

The camera and deck ran on an AC power or on a heavy battery power (camera 60 minutes; deck 90, on battery power). Batteries were recharged at night while the producer viewed the day's footage via an appropriate RF converter (see Glossary) on a motel receiver.

Travel was made easy because the camera was designed to fit under the seat, while the rest of the equipment was in compliance with the 75-pound-per-parcel passenger carrier regulations.

Some of their footage was about patients who did their own light housekeeping in return for cheaper hospital rates, as well as church fund raisers looking for a diamond field in Arkansas. Unfortunately, the "Money News Inserts" was canceled but it did generate excitement between traditional and electronic media among independent producers and directors.

Video is opening up a variety of industrial, documentary, and commercial situations for independent producers and directors. *Vogue* magazine taped 23 designer showings in New York, with an edited running time of approximately 26 minutes each, and traded the tapes to local retail outlets in return for ad space. These tapes were used to promote the retailers' own markets, thus proving video as a powerful marketing tool in commercial situations.

Medical

In Iowa City, Iowa, videotapes were used as therapy for anxious patients in a dentist's waiting room. Movies in the waiting room were found to ease the tensions by researchers at the University of Iowa College of Dentistry.

Researchers found that new dental patients were less nervous watching videotapes while waiting for their appointments. Viewing a pleasant experience before seeing the dentist makes it a lot easier to handle.

Three groups were tested. One watched a travelogue about Iowa City; the other, a patient having a good experience during a routine visit to the dentist; the third was a control group. The study showed that the two groups that viewed films were less tense than the control group. The relationship between high anxiety and cancelation of dental appointments may be reduced with video serving as a diversion. The long-range outcome of this study might be waiting rooms equipped with viewing areas that will allow the patients to select their own tapes and help reduce the anxiety experienced by adults in a dental setting.

Appendix: Companies Distributing Prerecorded Programs

For the Video Consumer who wants to be kept aware of the video scene, I would highly recommend two important magazines that are currently out on the market. One is *Videography* magazine, published at 475 Park Ave South, New York, N.Y. 10016. This magazine will help inform the consumer of the latest equipment on the market, plus featuring stories on the VTR revolution around the world.

My other choice is *The Videophiles Newsletter,* which comes out of 2014 South Magnolia Drive, Tallahassee, Florida 32301. The newsletter deals with other video cassette collectors who want to swap or trade prerecorded video tapes. Occasionally, people that sell used equipment might find some great bargains here.

Niles Films Products, Inc.
1141 Mishawaka Ave.
South Bend, Indiana 46615

Spectrum Tapes
1201 N. Watson Road
Suite 142
Arlington, Texas 76011

Show Tapes Hollywood
P.O. Box 610-686
North Miami, Florida 33161

Spectra Vision
4035 Naco Perrin
Suite 104
San Antonio, Texas 78217

Blackhawk Films, Inc.
2895 Eastin Phelan Building
Davenport, Iowa 52808

Magnetic Video Corp.
Industrial Park
Farmington Hills, Michigan 48024

Cinema Concepts, Inc.
91 Main Street
Chester, Connecticut 06412

Syndicate Films, Inc.
7411 Hines Place
Suite 117
Dallas, Texas, 75235

Entertainment Enterprises International, Inc.
334 Minorca
Coral Gables, Florida 33134

Reel Images
456 Monroe Turnpike
Monroe, Connecticut 06468

Space Web
P.O. Box 952
Palisade Station
Fort Lee, New Jersey 07024

Video Presentations, Inc.
64 Godwin Ave.
Midland Park, New Jersey 07432

Video Tape Network
115 East 62nd Street
New York, New York 10021

Syndicate Films Inc.
7411 Hines Place
Dallas, Texas 75235

Meda
7243 Santa Monica Blvd.
Los Angeles, California 90040

Studio Films & Tape Inc.
6424 Santa Monica Blvd.
Hollywood, California 90038

National Video Marketing Inc.
1 East 57th Street
New York, New York 10036

Quality X Video Corp.
356 West 44th Street
New York, New York 10036

Red Fox Inc.
Route 209 East
Elizabethville, Pennsylvania 17023

Hollywood Film Exchange
1534 North Highland Ave.
Hollywood, California 90028

Glossary

AC–Alternating current; also called line voltage; used by electronic circuitry as a point of reference; the voltage of an alternating current rises to its maximum and then falls to a negative value equal in amplitude to that maximum; this alternation takes place 60 times per second.

Access Time–Time during tape playback between the moment information is called for and the moment it is delivered.

AC-to-DC Converter–An electronic unit, which rectifies the flow of alternating current to direct current.

Automatic Gain Control–An amplifier circuit designed to provide output levels within a very specific range no matter what the levels of input are.

Air–Refers to broadcasting and transmission; to record "off the air" is to record material being transmitted.

Air Check–Off-the-air tape copy of commercial or program for verification or competitive use.

ALC–Automatic level control or automatic light control.

Amperage Rating–The number of amperes needed by an electronic unit to function.

Ampere–A unit of measurement (commonly abbreviated to "amp") of the electrical current used by a particular circuit; equal to watts divided by volts.

Amphenol Connector–A brand of microphone cable connector.

Amplifier–An electronic circuit that strengthens electronic signals, within a certain amplitude/frequency range, as they emerge from the circuit.

Amplitude–Strength of an electronic signal as measured by the height of its waveform (as displayed on an oscilloscope, e.g.).

Aperture–Opening at the camera end of the lens, through

which the light image collected by the lens is allowed to pass to hit the vidicon target area.

Aperture Grille—A metal screen located just behind the inside of a TV display tube's screen surface; used to limit the points at which the electrons hit the phosphor coating of the screen.

Aspect Ratio—The proportions of the TV picture area; the aspect ratio of television is four units of width to every three units of height; this is expressed as a 3 × 4 or 3:4 aspect ratio.

Assembly Editing—A method of electronic editing; various taped segments are retaped in a predetermined sequence to produce a coherent whole.

Attenuate—To turn down or reduce the level of a signal.

Audio Cue—Identification of an event by the use of sound; a word, noise, or other sound which alerts those producing either an audio or video tape that something is about to happen. In video productions, certain words in the script are used as "cues" to denote shifts in action, camera position, microphones, or other technical matters; in electronic editing, audio cues are often used to signal edit points.

Audio Dub—To rerecord the audio portion of a video tape without disturbing the video portion of the signal; also, to make a copy of an audio tape.

Audio Head—The recording and/or playback unit on a video tape recorder; it receives the video signal, induces that signal onto the magnetic tape moving past it, and/or induces the signal from the tape for reproduction.

Audio-In—Input jack that delivers an audio signal to a particular piece of equipment; an input receptacle which receives an audio signal.

Audio Mixer—Electronic circuit capable of accepting a number of audio signals from various sources (microphones, tape decks, turntables) and combining them at relative signal levels to form one composite signal; a unit which "mixes" various sounds into one total sound and enables the operator to control the level of each sound source as they are combined to be recorded on a single audio track.

Audio-Out—Output jack which carries an audio signal from a particular piece of equipment; an output receptacle which delivers an audio signal.

Audio Track—That portion of the video tape on which audio information is recorded.

Automatic Level Control—When used to describe an audio signal control, means the same as automatic gain control.

Automatic Light Control—Automatically adjusts the target voltage to compensate for variations in light levels.

A/V—Industrial term for audio/visual; also audio/video.

Available Light—Source of lighting (both natural and artificial) present in the scene to be taped.

B/W—Black-and-white; monochrome.

Backing—The plastic ribbon, usually of mylar, from both audio and video tape, on which the iron oxide is deposited.

Backtiming—Reverse cueing technique for editing backspace, used in electronic editing.

Balum—An adapter to convert 300-ohm to 75-ohm antenna input for VCR's projection systems, etc.

Bandwidth—Number of frequencies contained in a designated channel.

Barrel Distortion—Distortion of a scene by a wide-angle lens; everything looks rounded and out of proportion around the edges of the scene when objects are very close to the lens.

Beam—A semi-coherent flow of electrons.

Beam Adjustment—A control on vidicon cameras which regulates the amount of current flowing in the beam.

Beam-Splitting System—Method of dividing up the color components of the image so that it can be cast on more than one vidicon target area. Beam-splitting systems are used in two-, three-, and four-tube color cameras.

Beta—Half-inch video cassette format, developed by Sony.

Biconcave—A lens configuration in which the lens element has an inward curve on both sides.

Biconvex—A lens configuration in which the lens element has an outward curve on both sides; a magnifying glass is the most common example of a biconvex lens.

Binder—Chemical adhesive used to bind iron oxide particles to video tape backing material.

Black—"Going to black" or "fading to black" figuratively means to bring down the curtain on a scene or an act.

Black Clipping—A video control circuit found in cameras and in VTRs that regulates and contains the black level of the video signal so that it does not disturb or appear in the sync portion of the signal.

Black Level—Minimal television voltage signal establishing blackness of transmitted image; the bottom level of the picture signal, below which are the sync, blanking, and other control signals that do not appear as picture information.

Blanking—Suppression; the process (and the period of time which that process takes) during the scanning of the raster area when the beam is shut off. Line blanking takes place when the beam is returning from the end of one line to begin another; similarly, the process and the period of time in which the beam finishes scanning one field and retraces its path to the top of the raster are to begin scanning the next field, called field blanking.

Boost—To turn up; to increase in volume; to make a signal stronger.

Break-Up—Picture distortion that results in a totally indecipherable image that only lasts for a few seconds.

Brightness Ratio—An indication, expressed as a ratio, of the difference between the whitest and the blackest object in a scene; the range from brightest white to darkest black as it occurs in the scene being recorded. Too wide a brightness ratio can lead to an unacceptable contrast ratio when the scene is displayed on a TV screen.

Brightness Value—Luminance; the relative brightness of a particular object in a scene; the point on the gray scale at which the object is between absolute black and absolute white, either of which can be used as a point of reference to determine the brightness value of the object; essentially a relative determination made by the observer.

Burn In—Overbright images retained on the surface of TV

camera tube (removed by photographing a brightly lit white card).

Bus–One complete channel of an audio or video mixing system, including inputs, gain controls, and an output; two or more buses are required for video signal switching.

Cable–A grouping of wires in a protective sheath used for the transmission of electrical power and/or signals.

Cable TV–The system whereby images are sent to TVs over coaxial cable (wire). There is usually a fee for this service.

Call Letters–FCC-assigned station designations.

Camera–The eye of the video system; capable of absorbing the light values of a scene and converting them to a corresponding series of electrical impulses, through the use of a cathode ray pickup tube such as the vidicon; a light-sensitive cathode ray tube (and its associated electronic circuitry and lens optics) which translates the light values of any scene it views into a set of voltage variations that can be used to transmit those light values to another cathode ray tube, such as that used for TV display.

Camera Chain–The camera and its associated electronics; used to describe the camera, camera control unit, and power supply of large studio camera systems.

Cameraman–The operator of a manually controlled television camera. Also *camera person, camera operator.*

Capacitor–A component used in audio and video circuits to store and release voltages within the circuit.

Capping Up–Covering the lens with its cap to protect the camera's vidicon from burns.

Capstan–A rotating shaft on the VTR which is turned by a motor and which, in turn, governs the speed of the tape as it proceeds from the supply to the take-up reel.

Capstan Servo Editing–Head override editing; a method of electronic editing in which a new video signal replaces an already existing signal without disrupting the picture.

Carrier Frequency–The particular wavelength of a certain frequency on which a signal is impressed for transmission in a coherent fashion to a receiver, where it is stripped of its carrier frequency, amplified, and reproduced; can apply to either audio or video.

Cartridge—Container for tape or film that has a single reel or endless loop.

Cassette—Container for tape or film that has two reels.

CATV—Community antenna television; a system whereby TV signals received from off-the-air broadcasts or otherwise generated are sent along a coaxial cable to TV receivers; originally developed in the 1940s as a method of providing TV reception in rural, mountainous areas; presently being developed commercially. Using present technology, more than eighty channels can be sent by a single cable to any TV set.

CCTV—Closed circuit television; any form of television that is locally originated and displayed; non-network, non-cable TV.

Channel—The space on the frequency waveband assigned to a particular television broadcast; width varies from country to country; in the U.S., about 6 MHz wide for each channel.

Character Generator—Device which electronically displays letters or numerals on a TV screen.

Chip Chart—Standard B/W gray scale test chart for TV camera alignment.

Chroma Keying—Electronic introduction of a color background into a scene; unlike B/W keying, color is present and color values can be adjusted by the operator of the keying unit.

Chromaticity—Subjective evaluation of the hue and saturation of an object.

Chrominance—Chroma; the hue and saturation of an object as distinct from the brightness value (luminance) of that object.

Chrominance Signal—That portion of the total video signal which contains the color information. Without the chrominance signal, the TV picture is received in B/W.

Clamper—Electronic circuit which sets the video level of a picture signal before the scanning of each line begins, to ensure that no spurious electronic noise is introduced into the picture signal from the electronics of the video equipment.

Clean Edit—An electronic edit of a video picture which has no noise, distortion, or other disruption when the signal changes from picture 1 to picture 2. In a clean edit the picture is instantly replaced by a subsequent picture.

Clipping—A circuit which removes the positive and negative overmodulations of a composite video signal so that they do not interfere with the picture of sync information.

Closed Circuit—A distribution system using wires or microwaves to connect receiving sets to transmission equipment. The classroom unit of camera and monitor and sometimes VTR is considered a closed circuit system.

C-Mount—A mounting plate for vidicon video cameras and 16-mm movie cameras which accepts a certain type of lens.

Coaxial Cable—A one-ground, one-conductor cable which can carry a wide range of frequencies as far as 1000 feet with little or no signal loss.

Coaxial Connector—A specially designed cable connector used in cable TV and other 75-ohm cable applications.

Color Background Generator—An electronic circuit used in chroma keying to produce a solid color background of any desired hue and saturation.

Color Bars—Society of Motion Pictures and Television Engineers (SMPTE) standard test bars, electronically generated bar-shaped video tape leader color pattern to match playback to original recording levels and phasing. Usually occupied by a 1000-Hz audio reference tone.

Color Camera—A video camera capable of changing both the brightness values (luminance) and the color values (hue and saturation, expressed as chrominance) of a scene into a series of electronic pulses.

Colorizer—Electronic circuitry used to generate a chrominance signal in relation to the gray values of a B/W video signal. Each graduation of gray from black-to-white is assigned a color value. The result is an artificially colored picture, which does not truly represent the scene.

Color Killer Circuit—An electronic circuit used in a VTR to suppress the 3.58-MHz color carrier frequency when a B/W video tape is being shown; the same circuit in a B/W VTR used to suppress the color carrier frequency when a color tape is being played back in B/W. Without a color killer, the color signal would appear in the displayed B/W picture as random noise.

Color Phase—The proper timing relationship within a color signal. Color is considered to be in phase when the hue is reproduced correctly on the screen.

Color Picture Tube—A cathode ray tube, the screen-end of which is capable of glowing, with the three primary television colors—red, blue, and green. Its cathodes produce three electronic beams (each corresponding to one of the three colors) and its raster area is coated with three different types of phosphor (each one again corresponding to one of the three colors).

Colorplexer—Encoder; electronic circuitry which processes three separate color signals—red, blue, and green—coming from the pickup tubes into one composite encoded color video signal.

Color Subcarrier—The carrier wave on which the color signal information is impressed; contains the burst signal and alternating phase color information. In the U.S. the color subcarrier is 3.5794 MHz (usually rounded to 3.58 MHz).

Color Sync—A control signal necessary for the operation of color cameras, SEGs, and monitors; consists of a 3.58-MHz burst (which sets the color phase and placement before each line is scanned) and a 3.58-MHz color subcarrier.

Compatible Color—A TV broadcast system which produces a color signal that can be received by either a B/W or color set. The luminance values and the chrominance values are broadcast as different portions of the total signal so that the luminance values are not dependent on the chrominance values for reproduction.

Component—Any portion of a total electronic system; a component can be the transistor on a circuit board.

Composite Master—Original program produced by editing various portions of other recordings onto a new reel of tape.

Composite Sync—Total sync system, containing both horizontal and vertical scan controls.

Composite Video Signal—Video signal containing both picture and sync information.

Compression–Audio term, comparable to "video clipping"; the automatic adjustment of volume variations to produce a nearly consistent sound level. Elimination of audio overmodulations produces a sound lacking in dynamics; it is never soft or loud, but always at the same level.

Conductor–Strand of wires capable of carrying an electronic signal along its length; a length of cable which conducts a signal from one point to another.

Contrast Range–The range of gray between the darkest and the lightest values in a scene; expressed as a ratio of light to dark, such as "20:1," and used to evaluate the picture on a TV screen.

Control Track–The lower portion along the length of a video tape on which sync control information is placed and used to control the recording or playing back of the video signal on a VTR.

Control Track Head–An audio head which records the control track information onto the tape during record and induces it off during playback.

Convergence–Three primary colors for TV (red, blue, and green), overlapping perfectly to form an ideal picture.

Corner Insert–A second video picture signal, inserted into an area of the first video picture signal. Corner inserts are achieved by stopping the horizontal and vertical scanning of the first picture in a predetermined area and inserting the second picture scanning portions into that area.

CPS–Cycles per second; current term is Hertz; the number of times per second an electronic event is repeated.

Crash Edit–A brute force electronic assemble edit which may leave a slight glitch or distortion at the edit point on playback.

Creepie-Peepie–Hand-held TV camera.

Crispness–The sharpness of outline of the television picture.

Crossfade–To fade out one video signal and fade in another as a simultaneous movement.

Cross Talk–A spillover of sound from a line to an adjacent line.

Cueing–Presetting a record, transcription, or a tape on the first playback machine, for immediate starting.

Cut—To replace one picture instantly with a second picture.
Cut Away—Videotape shot of interviewer which may be interspersed during editing process to avoid jump-cut editing of interviewer.
Cutoff—TV cutoff; section of transmitted image which is hidden from home viewers by the receiver's mask.
Cutting on the Action—A production or editing technique in which two events are set in contrast to each other; as event A is taking place on the screen, the camera switches to event B before event A has ended.
Cutting on the Reaction—A production or editing technique in which one event is followed by a scene which displays the results of that event; after event A has taken place, the camera cuts to event B to show the impact of event A on the plot or characters.
Cutting to Tighten—Editing procedure used to shorten a series of shots; used to eliminate excess footage and to produce a coherent whole.

dBm—dB rating which indicates the number of decibels a signal is above or below one milliwatt.
DC—Direct electrical current which, unlike AC, maintains a steady flow, does not reverse direction, and therefore, cannot be measured in cycles per second (Hz).
DC-to-AC Inverter—Electronic unit which converts direct current to alternating current; used with an AC-to-DC converter to change one AC standard to another; for instance, 220-volt 50-Hz current is changed to 12-volt DC, and then to 120-volt 60-Hz current through the inverter.
D-C Restoration—The ability of the TV set to react to changes in brightness as seen by a video camera.
Definition—The sharpness of a picture subjectively evaluated in terms of its resolution.
Deflection Coil—An electromagnetic coil wound around the cathode end of the cathode ray tube to produce a magnetic field which controls the movement of the electron beam.
Degauss—To demagnetize.
Demodulated—Description of a signal stripped of the carrier

frequency onto which it was modulated; a signal is demodulated after it has been broadcast but prior to its display.

Dew Control—Warning signal that indicates the presence of too much moisture for the safe operation of a video recorder.

Diaphragm—Element in a microphone which is vibrated by sound waves entering the mike. The vibrations of the diaphragm are converted into voltage variations to produce the audio signal.

Dichroic Daylight Conversion Filter—Lens filter that balances the color values of objects in direct sunlight so that they will match the values of scenes taped in artificial light.

Dielectric—An insulator placed between conductors to prevent them from touching and thus shorting out the signal being carried.

Digital—Translation of information providing easy signal regeneration with minimal noise, drift, or distortion.

DIN—Deutsche Industrie-Norm; German standard for electronic connections. DIN plugs can be three-, four-, five-, or six-pin plugs, depending on their use, although they all have the same outer diameters and appearance.

Dissolve—A slow crossfade; one picture gradually fades out, the next picture gradually fades in; can be written using the symbol "X" on script.

Distortion—Any electronic interference in video picture.

Distribution Amplifier—A piece of hardware that strengthens electronic video signals. This item is used when you take one VCR and play the signal from the tape on several TV monitors.

Dropout—Loss of a portion of the video picture signal caused by lack of iron oxide on that portion of the videotape, or by dirt or grease covering that portion of the tape.

Dropout Compensator—Circuitry that senses signal loss produced by dropout and substitutes missing information with signal from the preceding line; if one line drops out of a picture, it is filled in with the preceding line, resulting in no dropout on the screen.

Dubbing—Duplicating audio and/or video signal, such as on a composite master tape, to make additional tape copies. Dubbing puts the resulting copy or dub "one generation" away from the tape from which it was recorded. Can also refer to erasing an audio track and recording a new track in its place.

Dynamic Mike—A type of very sound-sensitive uni- or omnidirectional mike, which can stand rough handling.

Edit—To creatively alter original recorded filmed video tape material.

Edit Code—Time code–video tape retrieval code added to original recording, utilizing a time structure, hours, minutes, seconds, and/or 1/30's (frames) with visual readouts.

Editing Deck—A specially constructed video tape recorder which has, in addition to play and record circuitry, circuitry and controls to accomplish assembly or insert editing; an editing deck is used in conjunction with a second video tape recorder to record a master program tape from various tape-recorded segments being played back on the second VTR.

Effects Buttons—The pushbutton controls on a special effects generator which indicate the special effect (inserts, wipes, keying, etc.) available on that SEG (Special Effects Generation) and which are engaged when that effect is desired.

EIA—Electronic Industries Association; the people who determine recommended audio and video standards in the U.S.

EIAJ—Electronics Industries Association of Japan.

EIAJ Type No. 1 Recommended Color Standard—The compatible color standard established by the Electronic Industries Association of Japan; compatible with the EIAJ Type No. 1 Standard (see following entry) in that color tapes can be played back in B/W on EIAJ Type No. 1 B/W VTRs and B/W tapes can be played back in B/W on EIAJ Type No. 1.

EIAJ Type No. 1 Standard—Standard established by the Electronic Industries Association of Japan for half-inch helical scan video tape recorders.

EIA Sync—Also called EIA RS-170 sync; the standard waveform for broadcast equipment in the United States as established by the EIA.

Eight-Pin Connector—A type of jack commonly used for the VTR-to-monitor connection; provides a full set of audio and video connections–one ground and one lead each for audio-in, audio-out, video-in, and video-out.

Electret Condenser Mike–A very sensitive microphone requiring a DC power supply (usually a battery built into the mike).

Electron Gun–The assembly at the end of the cathode ray tube which produces the electron beam for scanning; consists of cathode, heater, and grids.

Electronic Editing–Repositioning video signal segments on a reel of videotape without physically cutting the tape; a re-recording of the video signal segments in different order.

Electronic Viewfinder–Viewfinder; viewfinder monitor. A small TV screen attached to the video camera which allows the operator to view a given scene exactly as it is being viewed by the camera.

Electrostatic Focus–Method of focusing the electron beam in a cathode ray tube without using electromagnetic coils around the diameter of the tube; its advantage over coils is that it improves the quality of the picture while requiring less power to operate.

Equalization–"Eq"; the normalization of an electronic signal, either audio or video; adding "eq" in audio means reshaping the frequency response to emphasize certain frequency ranges and eliminate others.

Equalizer–An audio or video circuit which provides equalization either automatically or manually.

Equalizing Amp–A video circuit which is preset to provide a certain equalization to the video signal.

Erase Head–Either an audio or video head which erases the signal on a videotape prior to the recording of a new signal on that tape.

E-to-E–Electronics to electronics; monitoring the output signal of a VTR while it is recording is an E-to-E process, since the signal monitored has not yet been recorded on tape; rather, a sample of the signal is being fed from the VTR directly to the monitor; with E-to-E it itsn't possible to be certain that the signal is being recorded on the tape.

ETV–Educational television.

Extenders–A lens accessory which lengthens the barrel and in so doing reduces the minimum focusing distance of the lens and increases the effective *f*-stop.

Fade—Fade-in being an image slowly coming out of back to fill image.

Fade—To vary the strength of a signal, as in fading in or fading out; can be used with reference to both audio and video.

Fader—A sliding potentiometer control with which an audio–video signal is faded.

Federal Communications Commission (FCC)—The federal agency responsible for making policy for and exercising control over all use of the air waves for broadcast purposes.

Field—The electronic signal corresponding to one passage over the raster area by the scanning spot; 262.5 lines (1/60 second) in the North American TV system; half a frame. Two fields interlace to make one frame.

Field Blanking—Field retrace period; that period of time during which the field scanning spot returns from the bottom to the top of the raster area; used to add control pulses to the video signal; occupies about 15–20 lines of the 525-line system.

Field Frequency—The number of fields scanned per second; sixty fields are scanned per second in the North American TV system, which is also used in certain other countries, including Canada, Mexico, and Japan.

Field Time Base—The pattern of a field and points at which a field changes; 60 Hz is the field time base of the North American TV system.

Fill-Light—The illumination of shadowy areas in a scene to establish the proper brightness ratio or contrast ratio within the scene.

Film Chain—A special motion picture projector combined with a video camera to turn movies into video.

Film Loop—A length of film with the ends spliced together so that it can be projected continuously; frequently used for special effects, such as rain, or smoke, which may be added to studio pictures by superimposition.

Film Transfer—High-quality motion picture film made from an original tape; kinescope recording.

Filter—Glass element whose ability to transmit light varies with its design; used to exclude certain wavelengths or types of light; sometimes needed for color and B/W recording.

First Generation–Original recording of a tape segment. The first time the signal is recorded on tape, that tape is called "first generation." Every subsequent recording of the already recorded segment will be a generation removed.

Flag–Black cloth hung at front of a light to protect the camera against stray light entering lens.

Flagging–Picture distortion caused by improper operation of VTR or VCR playback/monitor timing coordination.

Flag Light–Lighting a scene or setting with overall brightness without noticeable modeling or highlights.

Flying Spot Scanner–Film-to-video tape-transfer system utilizing an electronic shutter.

FM–Frequency modulated; frequency modulated RF; describes a signal that has been impressed on a radio carrier wave in such a manner that the carrier frequency changes in proportion to the original signal.

Focal Length–Distance between the optical center of a lens and the image plane (which, in the case of the video camera, is the pickup-tube target area). The distance is measured in millimeters and determines the angle of view of the lens.

Focus–The greatest possible resolution of an object, when the object seems to be sharp and defined; to bring an electron beam or a light ray to its minimum size.

Focus Coil–An electromagnetic coil surrounding the cathode ray tube and producing a magnetic field which controls the flow of the electron beam from cathode to target area so that it strikes target area as the smallest possible spot; works on the same principle as the deflection coil.

Focus Control–Focus ring; a calibrated lens control which focuses light rays going through that lens by moving the internal elements of the lens; the control which governs the focus coil in a video camera; the setting of the distance between vidicon and end of lens.

Fog Filter–Lens filter which lends the effect of a fog to a scene.

Follow Focus–The continual adjustment of the lens to keep an object in focus as either object, camera, or both are moving.

Foot-Candle–Amount of illumination received by a surface one foot from a lighted candle–metric equivalent unit is "lux."

Foot Lambert–One lumen or one foot candle of light covering a square-foot surface.

Format–As in "recording format"; ¾-inch U, ½-inch EIAJ are two record formats commonly used.

Frame–A complete TV picture composed of two fields; a total scanning of all 525 lines of the raster area; occurs every 1/30 second.

Frame Frequency–Number of frames occurring in a given period of time; usually 30 fps (frames per second) or one-half the field frequency of 60 Hz.

Framestore–Device that records and stores video information which it can retrieve in the form of a still frame picture; information is stored on a video disc or disc cassette.

Frequency–The number of times a signal vibrates each second; expressed as "cycles per second" or, more usually as "Hertz" (Hz).

Frequency Modulator–An electronic circuit which produces a carrier wave signal on which the audio or video signal is impressed.

Frequency Range–Frequency response; the width of frequencies from the highest to the lowest frequency which a piece of equipment is capable of handling without distortion. Most circuits have limits to their frequency ranges, beginning at a certain number of cycles per second and ending at a certain number of cycles per second; for instance, the frequency range of audible signals that the human being can hear is described as being between 20 and 20,000 Hz.

f-Stop–A calibrated control (f 1, f 2, f 3, f 4, etc.) that indicates the amount of light passing through a lens to the target area; a control which can be adjusted to vary the size of the lens iris. Higher number indicates smaller iris opening.

Gain–Amount of signal amplification; "turning up the gain" means increasing the strength of the signal; "turning down the gain," decreasing the strength; used in both audio and video to denote the relative strength of the signal in question.

Gamma–In video mathematics, the exponent of the power law used to approximate the curve between the output magnitude and the input magnitude of the signal actually used.

Gap–The small space in an audio or video head across the magnetic field produced when recording and induced on playback; the audio and video heads are small, horseshoe-shaped electromagnets and the gap is the space which this tape must contact for good recording/playback.

Generating Elements–The component which enables the sound waves entering the head of a microphone to be used to produce an electronic signal composed of voltage variations corresponding to the sound wave.

Generation–Each time a signal is recorded from a camera or other source (such as off-the-air broadcast) and then re-recorded from that original recording; first recording is said to be "first generation," first rerecording is said to be "second generation," and so on. The more generations, the more time-base error and the more noise on the video and audio tracks.

Genlock–Device synchronizing television signal sources; circuitry which locks the sync generator that is used to control cameras and the SEG to the sync signal from a prerecorded tape on a VTR so that the signal from that tape can be mixed through the SEG with live camera signals.

Glitch–Random television picture noise appearing as an ascending horizontal bar.

Gray Scale–The number of steps from black to white that a camera can resolve; how faithfully the light values of a scene (changes of brightness from black to white) can be rendered by any piece of electronic equipment, such as camera or monitor; may be used as equivalent to the contrast ratio.

Halo–The black area around a very intense source of light, as seen by the camera and monitor.

Head–May be either for audio or video; a small electromagnet which pulses magnetic signals onto a video tape moving past it or induces those signals off a recorded tape; audio heads are usually stationary, video heads move in reverse of the tape's direction in most VTRs.

Head Alignment–The positioning of the audio or video heads so that they describe the correct path at the correct angle across the video tape.

Head Clogging–Occurs when the gap of an audio or video head gets filled with dirt, grease, or oxide so that it can no longer record or play back a signal.

Head Drum Assembly–That portion of the VTR in which the video heads and their related mechanical and electronic controls are located. In helical scan the head drum assembly is the large circular unit around which the tape wraps as it passes the video heads.

Head Drum Servo–One method of controlling video tape during playback so that the video heads contact the tape with the proper timing (sync) to retrieve the information on the tape. The control track pulses are used to control the rotation of the video heads.

Head Drum Servo Editing–An inferior variation of editing. Instead of the tape being slowed or speeded up to maintain sync, the video heads are slowed or speeded up by the use of the head drum servo controls.

Headlife–Normal lifespan of a video head.

Heads Out–A reel of tape wound so that the beginning of the program is at the beginning of the tape.

Helical Scan Video Tape Recording–A type of video recording in which the video heads and the tape meet at an angle such that the resulting pattern on the tape is a long, diagonal series of tracks from the video heads, each diagonal stripe containing the full information for one field of video picture; named after the helical path.

Helical Wind–The screwlike configuration of the tape across the video heads, from the plane of the supply reel, through the plane on which the heads are rotating, to the plane of the take-up reel.

Hertz–International electronic term for cycles per second, usually abbreviated to Hz.

High Resolution–Descriptive of a camera or monitor capable of displaying a great number of scanning lines (1000–2000) which produce a picture that is very detailed, well defined, and sharp.

Hiss—A constant noise that is usually heard during tape playback on the high end of the frequency band.

Horizontal Resolution—A subjective evaluation of the number of vertical lines that can be seen in a horizontal direction; the better the horizontal resolution, the sharper and less blurry the picture.

Horizontal Sync—The sync pulses which control the horizontal line-by-line scanning of the target area by the electron beam.

Hot—Live wire; a conductor carrying a signal is said to be a hot conductor.

Howl—Positive feedback; in video, the wild, swirling effect which results when a camera is pointed into a monitor displaying the picture which that camera is producing.

Hue—Term used to describe the dominant wavelength of a color in a range that runs from red to yellow, to green, to blue, to violet, and back to red; all colors have a hue.

Hum—A low-frequency noise.

IATSE—International Alliance of Theatrical State Employees—set workers (many cameramen and operators also belong to this union known as IA).

IC—Integrated circuit; a very small electronic component, containing a photoetched, miniature circuit.

Image Enhancer—TV engineering accessory to improve apparent video resolution.

Image Retention—Lag; the vidicon pickup tube's tendency to retain an image on its target area after it has stopped scanning that image. Extreme image retention results in the image being burned into the target area.

Image Transform—Proprietary computerized high-quality video tape-to-film transfer system.

Impedance—The AC resistance of a component to the flow of a signal; expressed as high or low impedance, hi-Z or low-Z.

In-Line Color—Color TV tube system in which the three electron guns producing the primary colors of the color signal are next to each other in a straight line rather than in a triad, as has traditionally been the case in color TV manufacture.

Insert—A general term meaning the introduction of a secondary signal into an already existing picture; accomplished by keying, wiping, or crossfeeding.

Insert Edit—Insertion of a segment into an already recorded series of segments on a video tape; the inserted segment replaces one which must be of the exact length. Insert edits demand that the segment be edited in and then edited out at the end of the segment.

Insertion Loss—The loss of signal strength that occurs when a piece of video or audio equipment is added to the path of the signal flow from origin to display; can be corrected by using an amplifier to build up the signal strength again.

Instant Replay—Immediate playback of recorded material, either full speed or slow motion, usually of a live program.

Intercutting—A production technique in which a cut is made from a scene (long shot) to a detail of that scene (close-up) to clarify or emphasize a point.

Interface—To connect two or more components to each other so that the signal from one is supplied to the other. Feeding a signal between units that run on different standards is the most frequent form of interfacing, as in connecting a half-inch helical span VTR to a two-inch quadruplex machine.

Interlace—A scanning method is which the lines of two fields are combined into a frame in such a way that all the lines of each field are visible as part of the frame; the positioning of 262.5 lines from one field with 262.5 lines from the next field to form a full 525-line frame.

Intermittent Shutter—Rotating prismatic lens arrangement replacing normal camera shutter (for film chain application, a five-bladed intermittent shutter is generally used).

Inverter—DC-to-AC current converter, similar to a rectifier.

ips—Inches per second; the customary way of measuring tape speed on an audio or video tape recorder.

Iris—Iris diaphragm; an adjustable set of metal leaves over the aperture of a lens, used to control the amount of light passing through the lens. Iris openings are measured in *f*-stops.

Jack—Plug-in electronic connection or connector.

Jeep—To convert a TV receiver into a TV monitor or monitor/

receiver by rewiring the internal circuitry and adding input and output jacks for video and audio.

Joystick–Band control stick for operation of electronic equipment such as editing controllers and video switchers.

Jump Cut–Bad or jagged edit of tape or film.

Jumper–A short length of wire used to make a temporary electric hookup, of AC power, video or audio.

Junction Box–Portable terminal box for AC power; also portable terminal for cable interfacing.

K–Kelvin temperature–used for measurement of light source color temperature; 0°K is -275°C.

Kelvin–Also expressed as Kelvins or K, the unit of measurement of the temperature of light in color recording; light temperature affects the color values of the lights and the scene that they illuminate.

Keying–Keyed insert; inlay insert; one video signal being controlled by the waveform of a second video signal when they are combined to form a composite picture. The signal from source 1 fills in the scanning lines of the total picture of source 2 at the points where the picture goes above a certain preset gray level.

Keylight–The spotlight or main light on a scene which emphasizes the important objects in that scene.

Kilohertz–The unit of frequency equal to 1000 Hertz.

Lag–"Ghost" image, retained when past action occurs in the presence of insufficient illumination; also, camera lag.

Lap–Lap dissolve: a cross-dissolve into new material while dissolving out of old material.

Lavalier Mike–Microphone worn around the neck and resting on the chest cavity. Lavaliers are small and unobtrusive.

Leader–Leader is often used at both the head end and tail end of a tape or between sections of raw unedited material.

Lens–A series of optical elements, contained within a barrel or tube, which collect and focus light.

Lens Cleaning Brush–A very fine brush specially made for cleaning a lens.

Lens Hood–Tunnel-shaped cover to keep ambient light off the face of the lens.

Lens Mount–Assembly on the front of the camera, to which the lens is attached.

Lens Paper–Paper specially made for cleaning lenses.

Lens Speed–Measurement of the ability of a particular lens to collect light.

Lens Turret–Rotating disc on the front of a camera which permits several lenses to be mounted on the camera at one time to facilitate rapid interchanging.

Level–Audio or video amplitude or intensity; also, as in "Give me a level," a test of same.

Lighting Ratio–The brightness level of the fill light compared to the brightness level of the key light, or the shadowy areas compared to the brightly lit areas, measured as a ratio determined by the f-stop of the lens; a 1:2 ratio means that key is one f-stop brighter than fill; 1:3, a stop and a half, and 1:4, two stops.

Light Level–The intensity of available light measured in foot candles or lux.

Limiting–Holding the strength of a signal to a predetermined level.

Line Count–The number of scanning lines actually used to carry the video picture signal.

Line Frequency–The number of lines scanned in one second; in the U.S. system it is 525 × 60, or a line frequency of 15.7 kHz.

Line-Matching Transformer–An audio device used to match the impedance of a microphone to the input impedance of a mixer, tape recorder, or amplifier; a device which changes the output impedance of a mike from low-Z to hi-Z, or vice versa.

Line of Sight–High-band transmission (TV or FM) to TV/radio receivers between antenna and the horizon line.

Line Period–The length of time it takes for a line to be scanned and then retraced to the point where scanning of the next line will begin.

Line Scanning–The path over the target area of the electron beam, as it moves from the left edge across the area.

Line Time Base—The control of the horizontal deflection of the scanning spot so that it starts to scan each new line at exactly the right moment.

Lip Sync—Abbreviated form of "lip synchronization"; recording of actors' dialogue at the same time that the program is being filmed.

Live Titles—Titles on cards for studio pickup, as opposed to slide or film titles.

Log—A listing of various items comprising a broadcast, as required by the FCC.

Long Lens—High-focal-length lens with a long barrel; performs similar function to the telephoto lens, without the advantage of that lens' shorter barrel.

Long Shot—A camera angle of view taken at a distance and including a great deal of the scene area.

Low Level Lighting—A scene illuminated with under-50-footcandles of light; often results in a poor signal-to-noise ratio and/or poor contrast ratio in the recorded picture.

Low Light Lag—A blurring, image-retention effect which occurs when a vidicon tube in operating in insufficient light.

Lumen—A measurement of light quantity, taken at the source of light against a predetermined constant. Lumens per square foot equals footcandles.

Luminance Signal—Black-to-white brightness values of a scene, which produce a B/W display picture.

Lux—The metric measurement of light quantity; taken at the surface which the light source is illuminating. One footcandle equals 10.76 lux.

Machine-To-Machine Edit—Transferring video and audio material from one recorder to player to another recorder, in which the edit is made by a simple assemble edit button and a glitch or distortion may appear at the edit point.

Macro Lens—A magnifying lens capable of focusing down to a few inches.

Magicam—Proprietary matting system for using small-scale model sets and matting in actors electronically to save on set construction costs.

Magnetic Tape Developer—Special chemical solution applied to the control track edge of videotape which makes control pulses visible to the eye and allows precise cutting of the tape between pulses; necessary for physical tape editing.

Main Title—Major explanation of the program theme, as opposed to "subtitle."

Master—Original completed videotape (or disc, or audio tape).

Master Monitor—High-quality monitor equipped with such facilities as picture focus, internal and external sync, and horizontal and vertical scanning controls.

Master Volume Control—An audio term, most often used with mixers and amplifiers to denote the final overall volume control of signal level.

Master VTR—When duplicating tapes, that deck which plays the original tape is called the master VTR; that which records the original signal on blank tape to produce the copy is called the slave VTR.

Match Cut—Editing in another camera's view of an identical moment in the recorded action.

Match Dissolve—Fading to or dissolving to an identical camera position.

Matching Transformer—A circuit which changes the impedance of a TV signal, often from 75 ohms to 300 ohms, for audio.

Matte—A film term sometimes used in video production work to denote a keyed effect, an insert of video signal information keyed from one source into a second video signal.

Medium Shot—Camera angle of view between close-up and long-shot; a view of the head and shoulders of a subject; as opposed to "head only" or "full body."

Memory—Magnetic information storage which is retrievable.

Meter—Unit consisting of a calibrated dial and swinging needle, which give a visual indication of the operation of the particular circuitry it is connected to.

Microsecond—One millionth of a second.

Microwave—A line-of-sight, cable-free transmission system for relaying broadcast signals with a wavelength of less than one meter, generally, for a radius between 8–16 km (5–10 miles).

Minicam—Describes lightweight, often self-contained portable-type camera.

Mini Plug—A female receptacle which accepts a mini jack; similar to a phone plug in design but much smaller; a plug introduced by Japanese electronic firms for use on miniaturized pieces of equipment.

Mistracking—Improper tape path and tape-to-head contact, resulting in bursts of noise appearing in the picture during display.

Mix—Session in recording studio, generally referred to as audio mix.

Mobile Unit—A location vehicle used for collecting or transmitting TV signals.

Mode—Electronic setting activating specific circuits in a system; e.g., record mode, playback mode.

Modulation—The process of adding audio or video signals to a predetermined carrier signal.

Moiré—Optical disturbance caused by interference between similar frequencies.

Monaural—Single sound source to both ears.

Monitor—TV set without broadcast receiving circuitry used to display directly the composite video signal from a camera, video tape recorder, or special effects generator.

Monitor/Receiver—A combination of monitor and TV receiver capable of accepting composite video signals directly from source or those video signals broadcast as RF; also capable of producing a composite video signal output from a broadcast input signal, allowing user to record "off-the-air."

Monochrome Signal—A black-to-white video signal containing only luminance information and capable of being received either by a B/W or color TV receiver and displayed as a B/W picture.

Montage—Visual blending of several scenes.

Multiplexer—An optical system allowing a number of film and slide projectors to feed video information into the same video camera.

NAB—National Association of Broadcasters, standard-setting and fraternal organization of the broadcasting industry.

NABET—National Association of Broadcast Employees and Technicians, a broadcast technicians' union.

NAEB—National Association of Educational Broadcasters.

Narrator—An on- or off-camera neutral performer commenting on the program's story or meaning.

NCTA—National Cable TV Association, the operators' membership association.

NET—National Educational Television.

Network—A group of affiliated broadcast stations such as ABC, CBS, or NBC in the U.S.

Neutral Density Filter—A filter placed over a lens to reduce the brightness of a scene without altering its colors.

Noise—Any unwanted signal present in the total signal; both an audio and video term to describe one signal interfering with another; usually created by some malfunction of either a component or circuitry which is part of the signal path; a signal inherent in certain audio or video components.

Normal Lens—Subjective evaluation of the angle of view of a lens; a normal lens is one which is neither wide angle nor telephoto.

NTSC—National Television Standards Committee; a broadcast engineering advisory group established in the 1940s that recommends standards to the FCC, for the 525-line 60-field system.

NTSC Color—Color standard used in the U.S. and set by the National Television System Committee; compatible color which can be received in B/W.

Octopus Cable—A cable which allows two pieces of video equipment with dissimilar jacks to be interfaced.

One-Tube Color Camera—A color-capable video camera that produces a color signal through the use of only one pickup tube.

On the Line—A term used in video production with a special effects generator to identify the signal which is leaving the SEG for broadcast or recording.

Optical Viewfinder—A mechanical device that enables the user to see how much picture the camera is taking in and thereby to frame the picture.

Oscilloscope—The oscilloscope is similar to a television set except that it shows waveforms instead of pictures.

Output—The terminal point of a unit of electronic equipment, from which the signal is taken.

Outtake—Taped segment not used in final edited version of a program.

Overscan—TV picture beyond area of normal screen size.

PAL—Phase Alternate Line; 625-line 50-field system, used in the U.K., Western Europe, Scandinavia, Australia, and South Africa.

Pan—To follow action by swinging camera left or right.

Passive Mixer—An audio mixer containing no active electronic components or circuitry; usually, an inexpensive audio mixer capable of combining and regulating the level of various signals but producing a loss in the strength of those signals, since they are not amplified with the mixer.

Patch Cord—Short cable with male connectors at both end.

Patching—The act of connecting two components to each other with a patch cord and/or patch panel.

Patch Panel—A plate on which a number of female receptacles are mounted, each the termination of a different audio or video signal; used with patch cords to make secure but temporary connections between components.

Patch Plug—Console-mounted female cable connection.

Peak-to-Peak Voltage—Total voltage produced by a signal, determined by adding together the positive and negative extremes to which the voltage modulates.

Peak White—Brightest, whitest portion of the picture signal, corresponding to the highest level the signal attains.

Phase—The relative timing of a signal in relation to another signal; if both signals occur at the same instant, they are in phase.

Phono Plug—Variety of jack most often used with audio amplifiers. Also known as RCA plug.

Phosphor—A chemical coating used on the inside face of a cathode ray display tube. When hit by electrons, the phosphor glows with a strength in ratio to the strength of the electrons.

Photoconductor—Any unit which permits the flow of an electrical current corresponding to varying light input.

Photoflood—Self-contained light bulb which gives a very bright intense light without the use of external lenses or lamp housings.

Picture Area—Area of TV screen containing the video picture.

Picture Locking—Synchronizing the picture signal; sync controls on a picture.

Picture Signal—Picture information part of the composite video signal; the portion of the video signal above the pedestal.

Picture Tube—Cathode ray tube designed to display the video picture signal.

Pinch Roller—A rubber roller which "pinches" or presses the video tape to the capstan. Together with the capstan, the pinch roller pulls the tape through the tape path on the video tape recorder.

Playback—Function which induces the magnetic pattern on a videotape from that tape into the circuitry of a video tape recorder, in order to reconstruct the composite video signal for display.

Playback Amplifier—In audio, a circuit which amplifies the audio signal prior to its being reproduced through a speaker; in video, a circuit which amplifies the video signal in the VTR prior to its being supplied to a monitor.

Playback Head—Audio or video head used to obtain a signal from the videotape. Some heads are capable of playback and record functions, others of playback only; the video heads of most helical scan VTR's serve as both record and playback heads.

Plumbicon—Trade name of N.V. "Phillips" special lead oxide tube, which is more sensitive than a vidicon; used in some color cameras.

Plus Diopter—Special lens accessory which fits over a camera lens to make the lens capable of extreme close-ups.

Polarity—Positive or negative orientation of a signal; in video, the polarity of the picture is black/negative, white/positive; reversed polarity would result in a negative picture.

Polarizing Filter—Special filter with polarizing properties; a filter which, when placed over the lens, can be rotated so that it cuts down the amount of reflected light coming into the lens.

Pop Filter—Sponge rubber or plastic foam cap placed over the end of a microphone to reduce sibilance, breathy sounds, popping *p*'s and *b*'s, and other unwanted vocal effects.

Power Pack—Rechargeable battery power supply or belt.

Preamplifier—Electronic circuit which maintains or establishes an audio or video signal at a predetermined signal strength, prior to that signal being amplified for reproduction through a monitor or speaker.

Preproduction—Covers all activity prior to actual taped production.

Preselector Box—Special tuner supplied by cable companies that adds the special cable channels to your TV set.

Preview—Monitoring of a video signal prior to its being processed through the SEG.

Primary Colors—Three colors used in color TV, no two of which can be combined to produce the third; red, green, and blue.

Processing Amplifier—Proc-amp, signal processor, video processor, helical scan processor. A unit inserted on the line between any two components through which a composite video signal travels; serves to stabilize the composite signal, regenerate the control pulses, and in certain models, change the gain and pedestal to improve contrast.

Projection Television—A combination of lenses and mirrors, which projects an enlarged television picture on a screen to attain a larger display area than a cathode ray tube is capable of.

Pulse—The variation of a constant signal for a certain period of time.

Pulse Distribution Amplifier—An amplifier designed to boost the strength of the sync as well as other control signals to the proper level for distribution to a number of cameras, special effects generators, and the like.

Quad—Quadruplex; four-head recording system, such as 2 inch quad, which writes video information on successive vertical stripes

Quadruplex—A system of video tape recording using 2-inch tape and four rotating video heads. The heads pass the tape at an angle perpendicular to its path.

Quartz Bulb–A small lighting element which produces a great deal of light for a long period of time.

Quick Cuts–A series of quick edits with dissolves.

Rabbit Ears–Home-indoor-type TV antennas.

Rack–To pivot a camera lens turret; also, rack focus.

Radio Frequency–RF; the range of frequencies used to transmit electromagnetic waves; a broadcast of that frequency range assigned to a certain bandwidth of that frequency.

Random Access–Simple retrieving of stored magnetic information regardless of where it is located on the tape.

Raw Stock–Unrecorded video tape.

RBG Signal–Chrominance information: red, blue and green.

Readout–Visual display of stored information.

Real Time–Original time span, without compression or selective grouping.

Rear Screen–Projected image film or slide from behind a translucent screen.

Receiver–Electronic component capable of collecting radio frequency broadcasts and reproducing them in their original audio and/or video form.

Recording Amplifier–Amplifier used in a video tape recorder to set the level of the video signal prior to its being supplied to the video heads.

Registration–An adjustment associated with color TVs to ensure that the electron beams for the three primary colors of the phosphor screen are hitting the proper color dots/stripes; also, a similar adjustment of the tubes in color cameras.

Remote–Broadcasting originating outside the studio,

Remote Pickups–Events televised away from the studio by a mobile unit or by permanently installed equipment at the remote location.

Resolution–Subjective evaluation of the amount of detail in a picture.

Retake–To re-shoot rejected material.

Reverberation–Repetition of a sound after it has been originally produced, caused by sound waves bouncing off objects and surfaces and thus reaching the ear or microphone later than the orginal sound.

RF Adaptor–RF amplifier; RF modulator/converter. A unit which accepts the composite video signal to modulate a carrier frequency and produce a broadcast signal on a standard TV channel.

RGB–Primary colors: red, green, and blue.

Roll–Loss of vertical sync, causing the picture to move up or down the screen.

Roll-off–The preset attenuation of a predetermined range of bass frequencies, used by some microphone manufacturers on their mikes to reduce the proximity effect.

Rotary Erase Head–A set of heads on the rotating video head assembly which erases the video signal during recording and editing; usually positioned one scan line in front of the video heads; produces cleaner edits than a stationary erase head.

Rotary Idler–Stationary guide along the tape path.

Rough Edit–Rapid assembly of various segments in the order they will appear in the final program; not a finished master tape; not a clean edit.

Safe Area–Ninety percent of the TV screen, from the center of the screen; that area of the display screen (and therefore of the camera scanning area) which will reproduce on every TV screen, no matter how it is adjusted.

Safe Title Area–Eighty percent of the TV screen, from the center of the screen; that area of the display screen (and therefore of the camera scanning area) which will reproduce legible title credits no matter how it is adjusted.

Safety–Extra copy of a video tape, in case something happens to the original copy; usually a second generation copy, although on special effects generators with two program outputs it's possible to record two master tapes, one being the safety tape.

Saturation–One of the determinations of the color of an object or light; how vivid a color is; related to the strength of the chrominance signal.

Scanning–Action of the electron beam as it traces a pattern over the target area of the camera pickup tube in order to convert the light values of each spot on the area to a corresponding electric signal.

Scanning Spot—Point at which the electron beam strikes the target area.

Schmidt Optical System—An arrangement of lenses and mirrors in combination with a very bright cathode ray tube; used in video production systems.

Scope—Cathode-ray-tube device for testing electronic signals.

SECAM—Séquentiel Couleur à Mémoire: sequential color memory; a color TV system developed by the French, which differs greatly from both the NTSC and PAL color systems; used in France, USSR, and Eastern Europe; 625 lines, 50 fields.

SEG—Special Effects Generator; see entry.

Separate Mesh—A mesh screen located in certain pickup tubes, which helps control the path of the electron beam from cathode to target area, thus improving the scanning process and the resulting picture.

SFX—Sound effects; music or sounds used as background for a production.

Shadow Mask Color Tube—RCA-developed dot matrix color picture tube; a color tube equipped with a metal sheet with half a million small holes in it. The metal is placed between the electron guns, which beam the picture signal, and the phosphor-coated screen.

Shoot—To videotape or film a production.

Shooting Ratio—Amount of tape recorded as opposed to the amount of tape actually used in the final, edited program.

Slave VTR—Video tape recorder used to record a copy of a video tape from another (master) video tape recorder.

Slow-mo—To slow down below normal tape speed.

Smear—Video picture in which objects are blurred at the edges and seem to be running or bleeding beyond the edges.

Software—Vehicle that can carry information or already has information on it for playback.

Sound Wave—Air (or other medium) set in motion by any physical force or entity. This motion is a vibration of a certain rate and strength which make it unique. The rate of vibration is measured in Hertz (Hz), or cycles per second; the intensity of the vibration is measured in decibels (dB).

Special Effects Generator (SEG)—Unit used in video production to mix, switch, and otherwise process various video signals to create a final signal, known as the program signal.

Spillover—Loud sound volume causing the VU meter needle to go above the 100 mark on the scale; the leakage of one signal from one line to an adjacent line; the leakage of a signal from one layer of recorded tape to an adjacent layer.

Spindle—Rotating shaft in tape transport system.

Split Screen—A special effect utilizing two or more cameras so that two or more scenes are visible simultaneously on each part of the screen.

Splitter—A device used to split one input into two outputs.

Spotlight—Lighting unit whose light can be focused into a beam and directed at a particular object or part of a scene.

Star Filter—Special filter to produce crosshatch or star effect by picking up light on object being shot.

Start Mark—Sync indication in either audio or video track or at the head of a film.

Step-Down Transformer—An electronic circuit which can change electric current from one voltage to another.

Still Frame—An individual frame of tape being held as a continuous shot.

Stock Shot—A piece of film or still photograph or stock tape from an archive, library, or collection licensed for reuse.

Streaking—When objects in a scene bleed beyond their edge.

Stretch—Slow down the action or add program material to fill in the allotted time.

Stripe Filter—A chrominance tube system is which the target area of the tube is divided into sequential stripes for red, blue, green, and yellow, and can therefore derive a color signal by using only one pickup tube.

Subcarrier Frequency—The frequency at which color information is modulated in the color TV system; in the U.S. it's 3.58 MHz.

Supply Reel—The reel on the VTR which contains blank tape or a recorded program prior to its being run through the VTR.

Switch—To cut from one camera to another (synonyms: cut, take, intercut).

Switcher—Term often used to describe a special effects generator; a unit which allows the operator to switch between video camera signals.

Switcher-Fader—A switcher with an added device used to fade an image out or in; fade from one camera image to another, or superimpose two images.

Sync—Synchronize; refers to various drive pulses, both horizontal and vertical, scanning procedures of the video picture signal from camera to display.

Sync Generator—A pulse generator which produces the sync signals necessary to integrate the functioning of various pieces of video equipment in relation to each other and the video signal.

T-Connection—A T-shaped three-way cable connector for distributing an incoming signal two outgoing ways.

Tails Out—A tape that has been played but not rewound.

Take-up Reel—The reel on which the tape is collected during recording or playback.

Tape—A medium capable of storing an electronic signal and consisting of backing, binder, and iron oxide coating. The orientation of the iron oxide determines whether the tape can be used for helical scan video recording.

Tape Path—Circuit followed by the tape from supply to take-up reel: past the erase head, video heads, audio/control track head, and between capstan and pinch roller; standardized on half-inch (12.7 mm) machines by the EIAJ.

Tape Production Circuit—Circuit which detects a spill or tear in videotape during playback or record and shuts down the recorder or player to prevent tape or machine damage. Available in some VTRs and VCRs.

Tape Tension Guide—The first guide off the supply reel, adjusted to maintain proper skew.

Tape Transport—Those mechanical components of the video tape recorder which move the tape from supply reel to take-up reel and back.

Target Area—Face of the vidicon tube or other camera cathode ray pickup tube. This area (opposite the cathode heater) is where the image formed by the lens is transformed into an electronic signal. On the outside face of the tube the image from the lens is focused; and on the inside, this image is "read" by the electronic scanning beam. A circuit is completed at each point at which the beam strikes the target, and because a voltage is being applied to the target area, a certain resistance results, which gives a voltage variation or video signal.

Telephoto Lens–Lens with a large focal length.

Television Receiver–TV set; capable of sensing and receiving broadcast video signals, stripping them from their carrier frequencies, and producing them as a light-image picture on the face of a cathode ray display tube.

Tension–The pull of the capstan assembly on the videotape to keep it against the video head drum assembly; used in conjunction with the skew control to keep tape properly in path.

Termination–The insertion of a load at the end of a line carrying a signal; a video terminator is a 75-ohm resistor placed at the end of a line to keep the signal from bouncing back along the line; 600 ohms is commonly used to terminate an audio line.

Test Pattern–Optical guide for TV camera reference alignment.

Three-Tube Color Camera–Color-capable camera which produces a color signal through the use of three pickup tubes, each assigned to one of the primary colors. An early stage in the development of the color video camera, introduced by RCA in 1940. Standard of the broadcast industry today.

Through-the-Lens VTR–The use of a small B/W video camera cabled to a monitor and/or recorder during filming with a motion picture camera.

Tin–To coat the end of a cable wire with solder before making a solder connection, thus ensuring a more sound connection.

Title Crawl–Roller-like device used to "roll" credits or titles across TV screen; either mechanical or electronic crawls are available.

Tracking–Angle and speed at which the tape passes the video heads.

Transducer–Element of the microphone that changes the sound vibrations into electronic pulses.

Transmission Ability—An evaluation, usually expressed as a percentage, of the amount of light a filter will admit. For example, a 90-percent transmission filter will allow 90 percent of available light to pass through it, eliminating 10 percent.

Transmission Line—Wire conductor designed to carry electronic impulses from one place to another with a minimum pickup of outside interference.

Tripod—A three-legged stand on top of which a camera is mounted.

Tripod Dolly—Base of tripod with wheels.

Tripod Head—Top portion of a tripod, where its legs meet and the camera is mounted; friction or fluid-head tripod designs are available.

Tuner—Device that adjusts the frequency broadcast range received by the TV. All TVs have two tuners: a VHF and a UHF.

TV Storyboard—Sheets of paper with blank TV screens on them; used for sketching out the action of a program.

Two Shot—Television picture showing two performers or two objects of major interest.

Two-Thirds-Inch (17 mm) Vidicon—A vidicon with a target area two-thirds of an inch (17 mm) in diameter; the most commonly used vidicon in portable video cameras.

U-Format—Refers to ¾-inch U-format video cassette recorders and players; equivalent terms are U-VCR, U-MATIC, ¾-inch.

UHF—Acronym for ultra high frequency; 470–890 megahertz (channels 14–83). More limited area of coverage than VHF, with the same power.

UHF Connector—Standard type of video-in/video-out jack; commonly found on professional monitors; used to carry either composite or noncomposite video signal.

VCR—A video cassette recorder (or player); generally, a ¾-inch U-format video cassette recorder.

Vectorscope—Round (green) oscilloscope to align amplitude and phase of the three TV color signals (red, green, blue).

Vertical Blanking–Field blanking; the blanking of a signal during scanning; when the scanning spot is flying back from scanning one field to begin scanning the next field, at which time blanking and sync pulses are introduced to the signal.

Vertical Interval–Moment (measured in microseconds) during which the electron scanning beam returns to the top of the TV tube.

Vertical Interval Switching–Method of switching video signals in a special effects generator; this replacement of one signal by another takes place during the vertical retrace period.

Vertical Resolution–Number of horizontal lines on a test pattern that can be reproduced by a camera or monitor so that they are distinctly visible; the number of horizontal lines a piece of video equipment is capable of processing per field as picture information.

Vertical Retrace–Return of a scanning spot to the top of the target area to begin scanning a new field after completing its scan of the previous field; the time it takes for this to occur.

Vertical Sync–Sync pulses which control the vertical field-by-field scanning of the target area by the electron beam.

VHF–Acronym for very high frequency; broadcast frequency band containing channels 2 through 13 on most TVs.

VHS–Acronym for Video Home System, the half-inch video cassette format developed by Matsushita.

Video–Visual portion of a signal containing both sight and sound information.

Video Amplifier–Circuit which can increase the strength of a video signal sent through it.

Video Cartridge–Self-contained video module played on a specially designed video tape recorder. The cartridge contains one reel of video tape, which is fed out of the cartridge into the internals of the VTR and then rewound onto the cartridge after play.

Video Cassette–Self-contained video module played on a specially designed video tape recorder; similar in design to an audio cassette; houses two reels–supply and take-up–with tape running between them but connected to both.

Video Distribution Amplifier–Special amplifier for strengthening the video signal so that it can be supplied to a number of video monitors at the same time.

Video Frequency–Composite video signal unmodulated by radio carrier frequency.

Video Gain–Amplitude of the video signal; the control on a VTR which determines the "volume" level of the video signal.

Videography–Term used to describe videotaping in a photographic sense, that is, taking motion pictures with video equipment; also refers to the whole range of video applications

Video-In–Jack through which video signal is fed into a given component.

Video-Out–Jack from which a video signal is fed out of given component.

Video Tape Recorder–(VTR)–Electromechanical device capable of recording and reproducing an electronic signal that contains audio and video information.

Viewfinder–Electronic viewfinder on a video camera.

Volume–Amount of fullness of sound.

Volume Unit Meter–(VU meter)– Meter generally associated with the monitoring of the amplitude of a video or audio signal.

VTR–Video tape recorder. VTRs are open-reel machines that do not take video cassettes or cartridges.

Wall Current–117 volts, in the U.S.

Watt–A unit of electrical power; that amount of power required to maintain a current of one amp under the pressure of one volt.

Waveform Monitor–Special oscilloscope used to display the video waveform.

Wide-Angle Lens–A lens with a very short focal length.

Wide Open–Descriptive of a lens set at its lowest *f*-stop rating so that the iris is opened as wide as possible.

Wild Footage–Video tape recorded without audio for use as visual material in post-production.

Wild Track–Unsynchronized video audio track.

Wind Screen–Heavy foam rubber microphone cover, used outdoors to cut down on audible noise created by wind blowing across the top of the mike.

Wipe–Term used to describe SEG effect of replacing a portion of video.

Xenon–Quartz lamp using xenon gas, giving constant color temperature.

Y Signal–Signal that's transmitted in color television broadcasts to give brightness and detail information.

Zoom Lens–Lens whose angle of view can be changed without moving the camera.